Systematics, Ecology, and the Biodiversity Crisis

Systematics, Ecology, and the Biodiversity Crisis

EDITED BY

Niles Eldredge

Columbia University Press
NEW YORK

Columbia University Press
New York   Oxford
Copyright © 1992 Columbia University Press
All rights reserved

Library of Congress Cataloging-in-Publication Data

Systematics, ecology, and the biodiversity crisis / edited by Niles
  Eldredge.
        p.   cm.
  Includes bibliographical references and index.
  ISBN 0-231-07528-6 (alk. paper)
  1. Biological diversity.   2. Species.   3. Species diversity.
  4. Ecology.   I. Eldredge, Niles.
  QH313.S97   1992
  333.95'11—dc20   91-46924
                                                                                      CIP

Casebound editions of Columbia University Press books are Smyth-sewn and
printed on permanent and durable acid-free paper.

Printed in the United States of America

c10 9 8 7 6 5 4 3 2 1

Contents

Introduction: Systematics, Ecology, and the Biodiversity Crisis

Niles Eldredge

Biologists in increasing numbers are turning their attention to the biodiversity crisis. Some have become activists in various aspects of the conservation movement. Others have maintained a more traditional detachment, seeking in the biodiversity crisis an improvement in biological understanding as they analyze the root causes of ecosystem collapse in the modern biota, as well as in mass extinctions of the remote geological past. Whatever the level and nature of its involvement, the academic community of biology now sees the biodiversity crisis as a very real phenomenon meriting our closest scrutiny.

To the eyes of a systematist, it is striking that by far the greatest concentration of organismic and population-level biologists addressing issues of biodiversity has come from the general ranks of ecology. Perhaps this is only to be expected: conservationists are surely right in their perception, now amply corroborated by the work of conservation biologists, that it is habitat alteration—reduction and destruction—that lies at the heart of extinction phenomena. And habitats support complex cross-genealogical arrays of local populations of species: communities, or ecosystems, depending upon one's precise theoretical ecological perspective. Cross-genealogical associations of different species are inherently the subject matter of ecology, not of systematics.

Systematics takes as its subject species and monophyletic lineages of species. Diversity to a systematist is the number of species within a monophyletic taxon, rather than the number of species represented by local populations within an ecosystem. The two indices of diversity—indeed, the very *meaning* of the word *diversity*—are different in ecology and systematics. The mechanics of extinction may lie squarely within the province of ecology, but we *measure* extinction taxonomically, squarely within the realm of systematics.

Systematists tell other biologists (and the human world in general) what the components of the living world are: what species exist, how they are related to other species, and in what part of the world they live. We can know that a species (or subspecies, or, for that matter, a phylum) has become extinct only if we first know of its existence. And the generation and preservation of this knowledge are the prime and virtually exclusive province of the systematist. It may well be that the dynamics of extinction processes will prove to be exclusively in the domain of moment-by-moment interactive processes of matter-energy transfer: the realm of ecology. But the problems of extinction can be defined, recognized, measured, and assessed only through the tools of the systematist.

This book seeks to explore the relation of ecology and systematics in their disparate approaches to problems of biological diversity. In particular, it explores various facets of systematics in relation to biodiversity in a frank attempt to heighten awareness of the real and potential importance that systematics has for understanding, and perhaps even helping to alleviate, the present-day biodiversity crisis. The book is an outgrowth of a symposium held at the American Museum of Natural History in March 1990 entitled "The Role of Museums in the Biodiversity Crisis." Participants subsequently modified their papers for publication, and a number of additional contributors were sought to round out the examination of the role not only of museums (and zoos and related institutions) but also of systematics more generally as it stands in relation to the biodiversity crisis. The book is to be viewed as filling a gap in the spectrum of other multiauthored volumes on the biodiversity crisis that have appeared in recent years (e.g., Soulé 1986; Wilson 1988; Western and Pearl 1989)—all of which have concentrated on ecological rather than systematics themes.

The majority of this book's contributors are members of the curatorial staff of the American Museum of Natural History. Each is a practicing systematist (as are most of the book's other contributors), though some explore additional avenues, including ecological approaches, in their biological research efforts. The aim was to stimulate a cross section of the systematists of one of the world's leading research institutions in systematics to explore the relation of their discipline to the biodiversity crisis. The result in no way reflects, of course, any overall single consensus on the role of systematics, or even museums per se, in the biodiversity crisis. No single institution, if it is any good at all, ever has unanimity and consensus among it staff. What has emerged, instead, is an intriguing array of connections drawn between systematics and the biodiversity crisis, from the minds of some of the finest systematists in the profession today.

The book's opening contributions look at the various disparate connotations of the very term *diversity*. Teasing apart ecological from systematics and evolutionary notions of diversity should help clear up some basic confusions—and ultimately help the analysis of causality underlying extinction episodes. Some of the early papers look at both ecological and systematics diversity, whereas others focus more on one or the other.

Later papers focus more explicitly on the roles of systematists and their institutions in confronting the biodiversity crisis. Museums are libraries of (systematic) diversity, providing a baseline against which all measures (even estimates) of diversity fluctuations are made. The contributions range from the theoretical to discussions of pragmatic, hands-on approaches of museums and zoos to biological conservation.

It is my profound hope that the contributions of this book will serve both to enhance the general appreciation of the importance of systematics in the modern world and to raise the consciousness of fellow systematists everywhere on the importance of our participation in meeting the serious challenge of the modern biodiversity crisis.

ACKNOWLEDGMENTS

Financial support for the two-day scientific symposium "The Role of Museums in the Biodiversity Crisis" (March 15–16, 1990) came through the efforts of two American Museum of Natural History Trustees: Mrs. Anne Sidamon-Eristoff and Mrs. Julia Serena di Lapigio. Ms. Flo Stone provided great organizational assistance, as did Dr. Michael Smith. Special thanks are due to Mr. Nathanial Johnson, Special Programs Coordinator of the Department of Education, the American Museum of Natural History; it was through his herculean organizational efforts that our plans coalesced and the symposium successfully took place.

REFERENCES

Soulé, M., ed. 1986. *Conservation Biology: The Science of Scarcity and Diversity.* Sunderland, Mass.: Sinauer Associates.
Western, David and Mary Pearl, eds. 1989. *Conservation for the Twenty-first Century.* New York: Oxford University Press.
Wilson, E. O., ed. 1988. *Biodiversity.* Washington, D.C.: National Academy Press.

Systematics, Ecology, and the Biodiversity Crisis

1 : Where the Twain Meet: Causal Intersections Between the Genealogical and Ecological Realms

Niles Eldredge

The relation between ecology and evolution has been undergoing close scrutiny in recent years. For example, some evolutionary biologists (e.g., Dawkins 1976) have attempted to subsume cross-genealogical phenomena, including (at least by implication, if not by specific detail) the dynamic structuring of ecosystems, under the general reductive evolutionary rubric of fitness maximization: both social and, more straightforwardly, purely economic systems are said to reflect, at base, the competitive struggle for reproductive success. In my view (Eldredge and Grene 1992, Eldredge in press), utilization of a dynamic (natural selection) that accounts for intergenerational stasis and change in gene frequencies as a description of the mechanics of cohesion of functional systems—especially patently cross-genealogical, economic systems—results in a rather distorted picture of biotic nature.

I take a somewhat different approach to looking for the connections between economic and genealogical systems in nature (Eldredge and Salthe 1984; Eldredge 1985, 1986, 1989; see Salthe 1985). It is an approach that sees economic and genealogical systems as basically independent but indeed connected—though connected in such a fashion that neither one subsumes, or emerges as somehow more fundamental, than the other. Before we can analyze the intersections and interconnections between ecology and evolution, we must first realize how separate, in biotic nature, these two sorts of processes really are. There is an underlying ontological reason why ecology and evolution have traditionally been rather separate areas of biological inquiry, reflecting an apartness of these two realms in the actual organization of nature.

I first develop why I think evolution and systematics (i.e., genealogical processes in general) are largely and intrinsically divorced from ecological process, before I begin to look for structural and functional (causal) links between the two—and indicate ways in which I believe understanding of the various measures of diversity in the two disparate areas may in fact share some underlying causality. These issues are critical, not just for purely intellectual reasons, but also for the implications they may have for conservation biology.

Most biologists agree that biodiversity crises—meaning realized and impending extinction—are fundamentally a cross-genealogical phenomenon. Extinction is a reflection of habitat degradation. Consequently, most practical responses to biodiversity problems focus on the preservation, even the enhancement, and certainly the protection, of areal expanses of various sorts of habitats. Yet we *measure* extinction by the number of *taxa* that have been affected: the number of species currently going extinct, or, in the case of mass extinctions of the past, the numbers of genera, families, orders, even classes and putatively phyla, that have disappeared.

Small wonder, then, that extinction may seem to some as a directly genealogical issue. Causes of specific taxon extinction (such as that of the dinosaurs, for example) are still being sought occasionally, outside the general context of ecosystems—while it is forgotten that many taxa other than dinosaurs disappeared at about the same time. Nor is this phenomenon restricted to paleontology. In February 1990 the National Research Council hosted a meeting looking into the worldwide, simultaneous decline of many species of frogs and salamanders, with the view to pinpointing common causes pertaining explicitly to Lissamphibia wherever they occur, irrespective of habitat. Interestingly, and not surprisingly, no single environmental factor, on a worldwide scale, could be identified that might account for the general phenomenon of decline in numbers of many different lissamphibian species.

Thus we have several kinds of phenomena masquerading under the general rubric of *biodiversity*. There are (1) *ecological diversity:* the number of different sorts of organisms present in a local ecosystem; (2) *genealogical diversity:* the number of taxa within a monophyletic clade—for example, the number of species within a family; (3) *phenotypic diversity,* or the amount of variation (or differentiation) within or among populations, or within species or still larger taxa. Gould (1989) calls this latter category of diversity "disparity." The focus of evolutionary theory ever since Darwin (1859), the generation of patterns of stasis and change in organismic phenotypic diversity, lies outside my present concerns. For the rest of this paper, I will be concerned strictly with ecological and genealogical diversity and their interrelationships.

Ecological and genealogical diversity are obviously not the same thing at all. Yet we sometimes behave as though they might be, as when ecologists list the number of "species" present in a local ecosystem (when there are only local populations, each a part of a species, present), or when ecological diversity-regulating theory is used (sometimes uncritically) to explain genealogical diversity patterns.

The Consequences of Organismic Economic and Reproductive Activity

It is apparent that there are direct biotic consequences of organismic activity (defined as the utilization of organismic phenotypic properties in the broadest sense: morphological features, physiologies, and behaviors). Moreover, there are two, and only two, classes, or categories, of such organismic phenotypic properties, viz., those that pertain to: (1) economic activity—boiling down to matter-energy transfer processes involved in the differentiation, growth, and maintenance of the soma; and (2) reproductive activity—the process of making more entities of like kind.

Merely by engaging in one activity or the other, organisms automatically find themselves as parts of larger scale biotic entities or *systems*. These are the *economic* and *genealogical* hierarchies, respectively (Eldredge and Salthe 1984; Eldredge 1985, 1986; Salthe 1985). The two hierarchies are fundamentally, intrinsically, and ontologically quite separate systems.

Organisms are simultaneously parts of both sorts of systems. In other words, any single organism (with some special exceptions) is always a part of a local economic system, as well as a local reproductive population. Because this is so, it is obvious that the two hierarchies converge at the organism level (a point to which I return below). But above the organism level, the two systems immediately begin to separate: sexually reproducing organisms, by the very nature of the process, are parts of local reproductive communities, or demes. The very process of sexual reproduction, in other words, creates a local aggregation within which mating and reproduction are carried out.

Simultaneously, at least for many metazoans and metaphytes, conspecifics interact in purely economic ways to form local economic aggregates—what Damuth (1985) has termed *avatars*. Avatars are local populations in an economic sense; demes are local populations in a reproductive sense. Often, no distinction is drawn between these two kinds of entities. For sessile benthic marine invertebrates, this may well make sense. But consider the composition of many fairly discrete local populations, which varies depending upon the time of year—meaning whether or not it is the reproductive season. Male elephants are loners through most of the year, while females and juveniles form herds. Ian Tattersall (personal communication) has recently reported that, in some species of the lemur genus *Propithecus*, local social aggrega-

tions, which would at first glance seem to be good candidates as both reproductive and economic entities, actually change composition radically once the
reproductive season arrives. Nonetheless, local aggregations of conspecifics,
in both economic and reproductive senses, result in groups that are sufficiently
similar to provide an additional potential point of intersection between the two
hierarchies—a point to which I return briefly below.

Yet, in the usual description of the two hierarchical systems, demes and
avatars are usually seen as the beginning of a complete split between the two
sorts of biotic systems; rather than seeing avatars and demes, when closely
similar in composition, as a point where the two hierarchies come together, I
see them representing the final vestiges of identity between the two systems.
By that I mean that avatars are parts of local economic systems—call them
simply "local ecosystems," and likewise, demes are parts of "species." And
here, at the level at which we see systems whose component parts are local
aggregates of conspecifics, we see the total dichotomy between genealogical
and economic systems.

Local ecosystems are patently cross-genealogical in composition. Avatars
are interacting as parts of definitive economic systems, with characteristic patterns of energy flow. Species, on the other hand, are reproductive communities. Following Paterson (e.g., 1985), I see species as the largest aggregation
of organisms to share reproductive adaptations (his "specific mate recognition
system"). By definition, they are "genealogically pure."

Parenthetically, I acknowledge here that there are several sets of problems
in delineating the components of the economic and genealogical hierarchies
that I have discussed elsewhere. For example, systematists often recognize
differentiation in economic phenotypic properties within reproductively defined "species"; especially if that differentiation can be resolved cladistically,
parts of one "species" may actually be shown to be more closely related to a
second species (that is "reproductively isolated"). All this means is that there
are difficulties, stemming from the difference between economic and reproductive characters, in deciding just what a "species" is. For present purposes,
this is irrelevant, for both sorts of species remain genealogically pure—compared with the cross-genealogical composition of ecosystems.

Further problems arise on the economic side, for not all ecologists agree
that ecosystems can be meaningfully delineated. Moreover, the concept of
ecosystem explicitly admits abiotic elements, which to some biologists obscures, or at least unduly complexifies, the purely biological aspect of such
economic systems. I have dealt with these and related problems elsewhere
(Eldredge 1985, 1989) and must simply pass them over here. For present purposes, the analysis is on sufficiently firm ground: economic behavior leads to
association of local avatars, which are parts of local economic systems. At
this level the concept of *niche* becomes meaningful. Avatars have niches—

not organisms, and certainly not species. These systems are inherently cross-genealogical. Organismic reproductive behavior, on the other hand, leads to the formation of demes (local populations in which reproduction occurs); demes are aggregated, regionally, into species—which are reproductive communities self-defined by the possession of the male/female specific mate recognition system (SMRS).

These two sorts of systems are inherently, intrinsically different: one is interactive, involving moment-by-moment energy flow. The other is *informational:* the elements of the genealogical hierarchy are, at base, packages of genetic information. It has been my contention that genealogical entities are not economic interactors (as in the statement that "species do *not* have niches"). Species, to me, are composed of a number of semiautonomous demes (Wright's "colonies") that are quasi-redundant (in terms of the genetic information they represent). Species appear to me to function solely as suppliers of genetic information (i.e., organisms); that is particularly apparent when, for example, after an oil spill, communities ("ecosystems") are refurbished from outlying demes.

To complete this brief characterization of the economic and genealogical hierarchies: species are parts of larger scale monophyletic (genealogically pure) taxa—the higher taxa of the Linnaean hierarchy. Local ecosystems, in contrast, are parts of larger, regional, and progressively interregional, networks of interacting economic systems. Thus economic biotic systems are inherently spatial—and though they persist for periods of time, their hallmark is definitely moment-by-moment interactions. It is the arena in which the game of life is largely played. Genealogical systems, in contrast, are the by-product of history that act as reservoirs of genetic information. It is fair to say, in concluding this brief run-through of the two systems, that they are radically different sorts of systems.

The Search for Connections: I—Organisms and Natural Selection

When we look for connections between the economic and genealogical hierarchies, quite naturally we look first at organisms themselves. As I have already mentioned, organisms clearly have a foot in both the genealogical and economic camps. The connection, as I see it, at the organism level is purely reciprocal—and is usually viewed strictly in the context of natural selection.

The competition that Darwin had in mind for his concept of *natural selection* is specifically and particularly *economic* competition: differential economic success, among organisms in a local population, will lead (all other things being equal) to differential reproductive success—on the assumption of course, of heritable variation. Under natural selection, reproductive processes (including the processes of mutation and recombination) present a pattern of variation (a spectrum of variation in economic traits) each generation. Those traits are utilized in an economic context; as a side effect of the variation in economic success resulting from this genealogically produced variation, there is in turn a bias in reproductive success, which determines relative frequencies of such genetically based traits in the next succeeding generation. Thus natural selection produces, each generation, no more and no less than a scorecard reflecting relative economic success in the previous generation. Natural selection is a passive—but nonetheless extremely important—ledger of immediate past economic success.

Current evolutionary theory does not see selection in quite this way. In particular, I am concerned with the current trend toward adding an "evolutionary perspective" to ecological discourse—a trend that has already reached something of an epiphany in the rapid development of sociobiology. Let me explain this briefly: In "evolutionary ecology," "sociobiology," and in the pure form of contemporary evolutionary theory that I have been characterizing recently as "ultradarwinism," "competition" generally means "competition for reproductive success." It is the goal of each organism to maximize its reproductive success—as witnessed, for example, in the game-theory-inspired "evolutionary stable strategies" of Maynard Smith (e.g., Maynard Smith, 1977).

Far from being a passive reflection of what works economically, economic (including social) systems are now commonly and routinely described and analyzed under the umbrella assumption that *they are organized around the principle that each and every organism is out to maximize the representation of its genes in the next generation.* Among other things, this viewpoint treats economic systems as if they really are all about the perpetuation of genetic information, instead of being primarily what they *seem* at first glance to be about: economics, i.e., simply living an economic existence, obtaining energy resources, surviving, maintaining the soma (and, to be sure, on occasion, reproducing).

In other words, it seems to me that modern evolutionary biology tends to go far too far in blurring the distinction between organismic economic and reproductive activities. In contrast, I prefer to see the two as separable but linked by an intimately reciprocal interconnection. There *is* interconnectedness between an organism's economic and its reproductive lives, at least when seen at the within-population level: reproduction provides new organisms each generation—in a population that is variable, yes, but reflects past economic success as well. Reciprocally, as a mere side effect of how well organisms fare relatively as they lead their economic lives, there will be a bias in reproductive success. Evolution is history—the track record left in the generation-by-generation historical sequence of changing gene frequencies.

But the reproductive *"needs"* of organisms (perceived of as needs only through a curiously extreme version of evolutionary theory) can by no means be taken as of such overwhelming importance as to form the very basis of the structure and function of the economic systems themselves: *that*, to my mind, is a distorted description of biological systems.

The Search for Connections: II—The Population Level

I have already mentioned that, to the extent that a deme and an avatar are one and the same entity (that is, composed of the same organisms), we have at this level a perfect connection between the ecological and genealogical systems. And I have also indicated that, in many instances where, at first glance, it might appear that demes and avatars really are just different aspects of the same entity, they nonetheless differ subtly (or even not so subtly) in composition.

The point is important in several respects. First, there is a purely evolutionary aspect here. Hull (1980) has provided a prescription for the evaluation of claims of selection at any level: there must be an element of "replication" (let us just say "reproduction") and "interaction," such that differential success arising from interaction biases reproductive success. We see how natural selection fits in immediately. But what of higher level selection?

I have been claiming in recent years (cf. Eldredge 1989) that, because species are purely genealogical entities, i.e., are not parts of larger economic systems, they are not themselves "interactors." This means, again following Hull's criteria for recognition of selection, that "species selection," whatever else it might be, cannot be said to be formally analogous with "natural selection." The claim that species themselves cannot be ecological interactors is based on my conviction that species are composed of a number of semiautonomous local populations (avatars in the economic arena—demes reproductively), each (as avatars) integrated into a different local ecosystem. There are no megaecosystems in which entire species can be said to play a single concerted role ("occupy a niche"—contra Mayr 1982). Only when species are reduced to (or start off as) single populations can an entire species be said to be a part of an economic system. It is parts of species that are parts of local ecosystems.

Again, the focus on evolutionary issues being maintained for the moment, the fact that parts of species can also be parts of local ecosystems (to the extent, that is, that demes and avatars converge on identity) implies that such parts are both interactors and reproducers. As such, they may satisfy, at least potentially, Hull's criteria for selection—in this context, a population-level selection. Social systems are an arresting case in point: Eldredge and Grene

(1992) see social systems expressly as hybrid sorts of systems, formed through intimate and intricate (and idiosyncratically different) modes of reintegration of organismic economic and reproductive activities. Thus social systems appear, on the face of it, to meld the sometimes separate avatar and deme into a single, cohesive entity. Yet I have already mentioned Tattersall's counterexample of apparently stable bands of *Propithecus* abruptly changing composition at the onset of the reproductive season: males depart for other bands, and new ones come in to mate with the band's females. Even social systems, it seems, are not always an effective blend of organismic reproductive and economic behaviors at the population level.

Yet here, surely (and see the compelling account of this phenomenon in Tattersall, this volume) is where theory and the concerns of practical conservation biology converge. Most conservationists have already come to realize that, the utility of focusing on particular endangered taxa (spotted owls, snail darters, etc.) notwithstanding, the effective approach is habitat conservation—for several reasons: many species are quite far flung, even if endangered. Even in the tropics, with geographic ranges of entire species much more restricted than in temperate areas, increase in environmental patchiness amounts to the same thing: one saves habitats, thereby one or more populations of many species—rather than try to target an entire, single species. This is the level at which niche theory becomes meaningful: it is also the level at which natural selection actively occurs. (The point is that, as a series of quasi-autonomous demes/avatars, rarely can an entire species be said to be evolving in a unitary fashion.) Thus it is appropriate, in the final section of this paper, to turn to a consideration of niche theory—and how it has been utilized to explain both ecological and genealogical diversity patterns.

Functional Connections: Avatars, Niches, and the Regulation of Both Genealogical and Ecological Diversity

In an early version of this discussion, in a chapter in Eldredge and Cracraft (1980), we made an explicit link between aspects of ecological niche occupancy, on the one hand, and patterns of morphological transformation on the other. Niche occupancy reflects the utilization of the economic adaptations of organisms; it also lies at the heart of most analyses of ecological diversity.

But the real nexus, one that I have pursued at length elsewhere (e.g., Eldredge 1979, 1989), is that niche theory—especially aspects of *relative niche width*—appears to be implicated with patterns of differential species production and survival,—in other words, speciation and species extinction—the causal elements underlying genealogical diversity patterns. But niche width is also very much a component of ecological diversity theory. Thus, in suggesting how considerations of niche width lead to understanding of the factors underlying both genealogical and ecological diversity patterns, we may as well be approaching an understanding of how such considerations, at the population level, actually have dual effects, simultaneously determining relative diversity patterns in different ecosystems and numbers of species within higher taxa.

But first, a caveat: Vrba (in press) has pointed out some of the conceptual confusion attendant in Stenseth and Maynard Smith (1984), in their consideration of Van Valen's (1973) Red Queen hypothesis. They are dealing with the local population level within ecosystems—which is appropriate; that is exactly the level at which the Red Queen imagery makes sense. However, they extrapolate their results to make conclusions about speciation and species extinction; all of a sudden, these avatars become species, and global conclusions about phylogenetic patterns are reached inappropriately.

The connections I am looking for are different: I am saying that differential patterns in ecological diversity (as, say, between tropics and higher latitudes) fall out of a dynamic (i.e., aspects of relative niche width) that also, in a genealogical context, governs (at least in part) rates of true speciation and species-level extinction. How might this work?

One pattern that is emerging as fairly general in systematics diversity data are sister groups where one clade is relatively unspeciose and generally plesiomorphic with respect to its proportionately much more speciose sister clade. There are many examples turning up—including a number to be found in the compendium of cases of "living fossils" (edited by Eldredge and Stanley 1984). An excellent example was first presented by Vrba (1980), in her analysis of evolutionary trends in the Alcelaphini—wildebeests, hartebeests, bonteboks, and the like. The sister group here are the impalas—Aepycerotini.

The two lineages apparently split in the Upper Miocene, some five million years ago. Because horn morphology fossilizes well and is implicated in the SMRS of African antelopes generally, it is possible to get an unusually accurate assessment of species numbers even with fossils. Vrba (1980, 1984) reports that there is one single, geographically widespread species of impala currently extant; there are seven species of alcelaphines. During the past five million years, there has been a minimum of twenty-five additional species of alcelaphines, whereas the record suggests that only one, or perhaps two, additional aepycerotine species existed—and never more than one at any one time.

The aepycerotines are species-poor and have changed relatively little since their inception. They have low extinction rates, as well—judged from their persistence in the fossil record. Alcelaphines, on the other hand, are species-rich, showing high rates of both speciation and extinction. They have accrued much anatomical specialization during the past five million years: wildebeests are highly derived antelopes.

This pattern of speciation, extinction, and morphological change correlates closely with aspects of niche breadth within these two clades. Impalas are notoriously eurytopic. They occur in a wide range of habitats, from open plains to wooded grasslands and even forests. They browse on leaves and graze on grass. Alcelaphines, in contrast, are strictly grazers and are more restricted in the habitats they can occupy.

What, then, are the causal pathways that link ecological niche width parameters with patterns of differential rates of speciation and extinction? I give just one version of how this connection might work—i.e., where broad-niched species (eurytopes) tend to speciate and undergo extinction at rather low rates, in contrast with their more narrow-niched breathren. I do not claim that the theory is correct in all its essentials—and I concede that there are variant versions of such theory that differ from my model in significant details. This should, then, be taken as an example of what the causal pathways might look like.

I have recently been looking at speciation (the derivation of one or more descendant reproductive communities from a preexisting reproductive community) as entailing, as a minimally sufficient condition, change in the SMRS. In other words, I follow Paterson (e.g., 1985) in seeing species as reproductive communities, which are aggregations of organisms with a shared fertilization system. Speciation must, minimally, entail disruption of the SMRS—however much economic adaptive change accompanies it.

Parenthetically, note how this model stands the conventional synthesis model on its head: it is the accrual of (generally) economic adaptive change that gradually leads to reproductive isolation in the latter model, but elements of the SMRS can be modified (by sexual selection for continued mate recognition in allopatry, for example) in Paterson's model—irrespective of the degree to which economic adaptive change is accruing.

At any rate, there is no necessary correlation between economic and reproductive adaptive change; we are all well aware of monomorphic vs. polymorphic species, and sibling species vs. highly differentiated sister species. There can be a high or low degree of within-species differentiation, and a high or low degree of among-species differentiation as well—or, to use Gould's (1989) term, *disparity.*

Nonetheless, little in the way of economic adaptive change seems to accrue in the absence of speciation (reproductive adaptive change). This is what Darwin meant by his expression that species are "permanent varieties." Within-species differentiation tends to be lost unless injected into the phylogenetic mainstream by speciation, i.e., "reproductive isolation." This, I claim, is an empirical generalization of the fossil record; most adaptive change appears to be associated with true cladogenesis, i.e., splitting of lineages. How does this work?

I was struck, a number of years ago, by Lewis' paper (1966), in which he painted a picture of new species of *Clarkia* arising in California every year—

a simple chromosomal event, occurring at a fairly high rate, rendered small populations "reproductively isolated" from the parental population. But there being no economic differentiation, the fate of such new species is in the vast majority of instances to be swamped. There is no way for such fledgling reproductive communities to gain an ecological foothold, without any economic differentiation from the parental species. Thus, it appears that new species have a higher probability of survival as fledglings the more differentiated they are economically from the parental species.

And this is where eurytopy/stenotopy, that is, aspects of niche width, enter in. Stenotopes within clades show a marked tendency to subdivide and specialize with respect to various environmental parameters, including, in Vrba's antelopes, energy resources. Consequently, stenotopic species within the same lineage are often sympatric. Not so with eurytopes, which, as broad-niched organisms, tend not to be sympatric with closely related, eurytopic relatives. The patterns certainly suggest competition: eurytopes are unable to withstand competition with close eurytopic relatives—whereas, through niche subdivision, stenotopes can; this is actually simply a restatement of the empirical patterns.

Let us assume that there is equal probability for SMRS disruption among eurytopes as stenotopes; eurytopes tend to be far-flung, widely distributed; stenotopes tend to occupy patchy distributions, according to the distributions of their preferred specialized (sub)habitats. Both kinds of organisms would seem to have about equal probability for "reproductive isolation" to be developed. I am simply suggesting that stenotopes might be more likely to produce reproductively isolated descendant taxa that demonstrate modification of niche parameter utilization than eurytopes are. Eurytopes beget eurytopes; stenotopes seem to beget more stenotopes.

Returning to the imagery of Lewis (1966): eurytopes, ecologically less distinct from parental species than stenotopes might likely be, have a lower probability of survival as fledgling species. Thus they will appear to "speciate" less frequently than stenotopes—when what is actually meant is that *successful* speciation, where the new reproductive communities survive and live on, is less likely for eurytopes than for stenotopes.

But stenotopes have been known for more than a century to have shorter temporal durations than eurytopes. What does this mean? Stenotopes, being narrow-niched and often, if not invariably, less widespread in distribution than close eurytopic relatives, show characteristically higher extinction rates. Eurytopes, less focused on specialized subhabitats, simply have a greater chance of survival. And the longer any single eurytopic species survives, the less chance that a new one will arise; survival of an ancestral eurytope tends to dampen the chances of new species to get going.

That, in a nutshell, is my argument linking eurytopy with low rates of speciation and extinction, and conversely, stenotopes with high rates of speciation and extinction. I note, once again, that there are other ways to develop such models to explain the same basic set of observations. Now, let us take a quick look at the ecological diversity side of the ledger.

Throughout the foregoing discussion, I have been using, rather loosely, the expressions "eurytope " and "stenotope" for *species*—ignoring my own injunction against sloppy usage; for it is not species that have niches, but avatars—parts of species that are themselves parts of local ecosystems. Species have such properties only in the sense that their component organisms are physiologically broadly or widely tolerant—and to the extent that their avatars develop relatively narrow or broad niches within ecosystems.

Niche width has, of course, figured heavily in ecological diversity theory. In focusing on "Rapoport's rule" (that higher latitude species tend to have greater latitudinal ranges than tropical species), Stevens (1989) chose a clever way to sneak up on the old problem of why there are more species represented in tropical than higher latitude ecosystems. The model looks not at resources, but simply at temperature and perhaps rainfall. But the argument is nonetheless the same in kind: high-latitude species must withstand far greater environmental changes in the course of a year than tropical species—which would, so to speak, be wasting their time being broadly adapted. Conversely, habitats are far more patchily distributed, as far as the specialized needs of the stenotopic tropical forms are concerned, than they are in higher latitudes. Entire species are consequently far more restricted in their areal distributions than taxa typically are in the higher latitudes. In a nutshell, Stevens thinks that much of the high diversity (numbers of avatars within a local habitat) in the tropics is the result of constant supply of invaders in suboptimal habitat—where it might even be true that they are not even reproducing.

What are the points in common in these models? You can pack more stenotopes into a higher level entity than you can eurytopes; this is true of avatars in ecosystems, species in regional biotas, and species within higher taxa. (This is not true, incidentally, of demes within species: the number of demes in a species has nothing to do with degree of economic differentiation within a species). Within a monophyletic taxon, we might expect a greater number of eurytopes in higher latitudes than in lower latitudes; in one sense, this is the empirical base of Rapoport's rule—but I have no idea whether, in other words, the generalization pertains to other aspects of niche width vis-à-vis energy resources. This leads to a further, rather obvious point: Rapoport's rule involves *taxa*—whereas the relative diversity of the tropics vs. higher latitudes is a classic question in cross-genealogical ecology; Stevens himself has already forged a conceptual link between systematics and ecological diversity patterns.

And, finally, we should always remember that we measure diversity patterns—of all sorts—in terms of species. But the dynamics of causality arise from lower levels: the organisms (which are simultaneously parts of economic and genealogical systems) and populations of conspecifics—the demes and avatars, the actual level where "ecological niche" appears to have its greatest meaning.

There is much conceptual confusion surrounding the relation between ecology and evolutionary biology. In particular, I have stressed how organismic economic behaviors lead to the development of large-scale economic systems that are, as a matter of ontological reality, quite divorced from the genealogical system that develops as a simple consequence of organismic (sexual) reproduction.

That being said, and appropriate pitfalls to the unwary being clarified (e.g., species do not have niches; monophyletic taxa do not occupy "adaptive zones," if what is meant is some higher-level analogue of "niche"), it is clear that there are connections aplenty to be made between ecological diversity patterns and those we see in systematics that are the outcome of the evolutionary process.

It is organisms, and, I think, especially populations (avatars in an economic sense; demes in a genealogical sense) that bridge the two otherwise separate worlds of economics and genealogy in biotic nature. I have developed, especially for genealogical systems, models utilizing niche width as an exemplar. Much remains to be done. The essential point is that niche-width considerations are real and have characteristic implications for long-term genealogical history (producing characteristic taxic diversity patterns) as well as for ecological diversity patterns. We need now to see what factors of niche utilization at the population level produce which characteristic genealogical patterns and which ecological patterns. There should be many interesting connections made in the coming years as we follow up these lines of thought.

REFERENCES

Damuth, J. 1985. Selection among "species": A formulation in terms of natural functional units. *Evolution* 39:1132–1146.

Darwin, C. 1859. *On the Origin of Species.* London: John Murray.

Dawkins, R. 1976. *The Selfish Gene.* New York and Oxford: Oxford University Press.

Eldredge, N. 1979. Alternative approaches to evolutionary theory. In J. H. Schwartz and H. B. Rollins, eds., *Models and Methodologies in Evolutionary Theory. Bulletin of Carnegie Museum of Natural History* 13:7–19.

Eldredge, N. 1985. *Unfinished Synthesis. Biological Hierarchies and Modern Evolutionary Thought.* New York: Oxford University Press.

Eldredge, N. 1986. Information, economics and evolution. *Annual Review of Ecology and Systematics,* 17:351–369.

Eldredge, N. 1989. *Macroevolutionary Dynamics. Species, Niches and Adaptive Peaks.* New York: McGraw-Hill.

Eldredge, N. In press. Ultradarwinian explanation and the biology of social systems. In E. L. Khalil and K. E. Boulding, eds., *Social and Natural Complexity.* London: Academic Press.

Eldredge, N. and J. Cracraft. 1980. *Phylogenetic Patterns and the Evolutionary Process. Method and Theory in Comparative Biology.* New York: Columbia University Press.

Eldredge, N. and M. Grene. 1992. *Interactions: The Biological Basis of Social Systems.* New York: Columbia University Press.

Eldredge, N. and S. N. Salthe. 1984. Hierarchy and evolution. *Oxford Surveys of Evolutionary Biology* 1:182–206.

Eldredge, N. and S. M. Stanley, eds. 1984. *Living Fossils.* New York: Springer Verlag.

Gould, S. J. 1989. *Wonderful Life: The Burgess Shale and the Nature of History.* New York: W. W. Norton.

Hull, D. L. 1980. Individuality and selection. *Annual Review of Ecology and Systematics* 11:311–332.

Lewis, H. 1966. Speciation in flowering plants. *Science* 152:167–172.

Maynard Smith, J. 1977. Parental investment—a prospective analysis. *Animal Behavior* 25:1–9.

Mayr, E. 1982. *The Growth of Biological Thought.* Cambridge, Mass.: Harvard University Press.

Paterson, H. E. H. 1985. The recognition concept of species. In E. S. Vrba, ed., *Species and Speciation. Transvaal Museum Monographs* 4:21–29.

Salthe, S. N. 1985. *Evolving Hierarchical Systems.* New York: Columbia University Press.

Stenseth, N. C. and J. Maynard Smith, 1984. Coevolution in ecosystems: Red Queen evolution or stasis? *Evolution* 38:870–880.

Stevens, G. 1989. The latitudinal gradient in geographical range: How so many species coexist in the tropics. *American Naturalist* 133:240–256.

Van Valen, L. 1973. A new evolutionary law. *Evolutionary Theory* 1:1–30.

Vrba, E. S. 1980. Evolution, species, and fossils: How does life evolve? *South African Journal of Science* 76:61–84.

Vrba, E. S. 1984. Evolutionary pattern and process in the sister-group Alcelaphini-Aepycerotini (Mammalia:Bovidae). In N. Eldredge and S. M. Stanley, eds., *Living Fossils,* pp. 62–79. New York: Springer-Verlag.

Vrba, E. S. In press. From ecology to macroevolution: The Red Queen, turnover pulses, and related topics. In N. C. Stenseth, ed., *Coevolution in Ecosystems and the Red Queen Hypothesis.* Cambridge, U.K.: Cambridge University Press.

2 : Patterns of Biodiversity

Norman I. Platnick

I consider here the relationships between various concepts and measures of "diversity" from the vantage point of a practicing systematist and biogeographer. Let me begin with a quote from Robert MacArthur's (1972:197) book *Geographical Ecology:*

> Perhaps the word "diversity" like many of the words in the early vocabulary of ecologists . . . should be eliminated from our vocabularies as doing more harm than good. To some people "diversity" means the number of species, to some it incorporates both the number and the evenness of their abundances, and to some it can be viewed as a vector with one component the number of species and the second component the evenness of abundances; to yet others it is best described by a relative abundance curve.

I argue that none of these ecological concepts of diversity is sufficient to allow us to answer the most pressing practical question posed by today's biodiversity crisis. This question is really biogeographic, namely: How do we determine where on this planet our limited resources, both financial and human, can best be placed so as to minimize the biotic impoverishment that we all realize is taking its irreversible toll, even as we speak?

We are, as has been widely documented, in a triage situation. So many different species and so many different habitats are being threatened in so many different countries, by so many different human economic needs, that it

15

is clearly impossible to preserve every species, and every habitat, that still exists today. Is there a rational way to decide where to concentrate our efforts?

To answer that question, I need to discuss some differences between general and comparative biology and show how these differences permeate discussions of biogeography and diversity. Then I show how fragmentary our knowledge of biogeography in the comparative sense really is and how severely it is affected by what I refer to as the megafauna bias and the boreal bias. And finally, I sketch how, with enough effort directed into inventories of threatened areas, into systematic analyses of the plant and animal groups occurring in those areas, and into biogeographic comparisons across those different groups, we might be able to set some priorities while there is still time for the decisions to make a difference.

First, the two biologies. In attempting to make sense of the panoply of subdisciplines and specialties that comprise modern biology, various routes have been taken. Perhaps the most common is the levels of organization approach, which generally leads to a division between subdisciplines operating at the level of organisms or above (organismic biology) and those concerned only with parts of organisms. Another approach is to separate evolutionary from structural and functional biology. But perhaps the most basic difference is simply between what has been called general and comparative biology (Nelson and Platnick 1981:4–5): "The general biologist usually works on a single species and regards it as an experimental tool, hoping to discover in it properties that may prove to be general. To such biologists, diversity is only a hindrance. . . . Commendable as such studies may be, they are most often thwarted. . . . Relatively few properties prove to be true for all living organisms." Comparative biologists are those who attempt to understand both the similarities and differences among organisms.

Even a cursory examination of the literature on biogeography quickly reveals the same dichotomy of approaches, this time focused on an ecosystem rather than species level. Indeed, in picking up a book or paper on biogeography for the first time, one scarcely knows which approach, or what mix of the two, it is likely to contain (Nelson 1978). MacArthur's book provides an excellent example of the general biologist's approach; his chapter on "Patterns of Species Diversity" (1972:169) begins as follows:

> There are more species of intertidal invertebrates on the coast of Washington than on the coast of New England, more species of birds breeding, and also more wintering, in forests than in fields, more species of diatoms in unpolluted streams than in polluted ones, more species of trees in eastern North America than in Europe, and more flies of the family Drosophilidae on Hawaii than anywhere else. There is an even more dramatic difference in the number of species in the tropics than in the temperate. . . . Will the explanation of these facts degenerate into a tedious set of case histories, or is there some common pattern running through them all?

I say more on the tropics versus the temperate below as well, but first another quote from the same chapter (1972:176):

> At the other extreme people often suggest we should explain the diversity of all living things, not just of trees, or birds, or butterflies. But to suggest this is not only a little masochistic because the counting job would be virtually impossible—it also misses the point. We are looking for general patterns, which we can hope to explain. There are many of them if we confine our attention to birds or butterflies, but no one has ever claimed to find a diversity pattern in which birds plus butterflies made more sense than either one alone.
>
> Hence, we use our naturalist's judgment to pick groups large enough for history to have played a minimal role but small enough so that patterns remains clearer.

Since MacArthur wrote those words, an alternative approach to historical biogeography (now usually called vicariance or cladistic biogeography; Nelson and Platnick 1981; Humphries and Parenti 1986) has suggested that there are indeed historical diversity patterns that allow us to make sense of the distributions, not only of birds and butterflies, but perhaps even of all groups of organisms.

Consider what sounds like a simple question, albeit one with considerable practical import: what parts of the world have the greatest biodiversity? The most obvious answer would be, simply, those parts of the world that house the largest number of different species. Ecological biogeographers have, in general, answered that the greatest diversity is in the tropics, rather than in the temperate parts of the world. They have produced counts of species to back that answer, often in the form of maps, such as MacArthur's (1972 fig. 8–8), showing estimated numbers of breeding land bird species in different parts of North America, going from low numbers in Alaska and Canada to high numbers in Central America. Similar maps are abundant in the literature on many groups, but what of their converse, i.e., maps placing the tropics at the top and south temperate regions of the world at the bottom and showing the number of species decreasing toward the south? Such inverted maps seem to be few and far between.

Biologists have done a fairly good job of explaining to the general public the severity of the biodiversity crisis and its importance. But those public presentations have almost invariably assumed that biodiversity in the tropics is vastly higher than anywhere in the temperate zones. Unfortunately, this is merely an assumption, rather than a demonstrated fact, and the ease with which it has been adopted and received seems to reflect two biases that have badly distorted our picture of the world—the boreal bias and the megafauna bias.

The boreal bias is simply the expectation that what is true for the Holarctic part of the world, namely, North America, Europe, and northern Asia, can be assumed to be true for the south temperate parts of the world as well. The boreal bias exists simply because the vast preponderance of biologists ever alive have worked within the Holarctic zone, especially in Europe and North America.

The megafauna bias is simply the expectation that what is true for vertebrates, especially big and obvious vertebrates like mammals and birds, can be assumed to be true for other groups of organisms as well. The megafauna bias exists simply because the majority of biologists who have ever worked on biogeographic questions have been specialists on vertebrates, especially mammals and birds. At the American Museum of Natural History, for example, 25% of our zoologists (8 of 32, to be precise) work primarily on mammals, and at least 20 of the 32 work exclusively on vertebrates. Yet vertebrates in total include less than 5% of the species on this planet, and mammals and birds together include far less than 1% of that total. Let me be blunt: speaking of biodiversity is essentially equivalent to speaking about arthropods. In terms of numbers of species, other animal and plant groups are just a gloss on the arthropod theme.

I do not dispute the claims that there are more species of mammals and birds in the tropics than in either the north or south temperate parts of the world. But I have had the good fortune to do most of my field work in south temperate areas and to work on spiders rather than vertebrates. From that vantage point, the world looks a little different. I have no quarrel with maps showing the numbers of species increasing from Canada to the United States to Mexico and Central America. That is generally true for spiders, and generally true moving south from Scandinavia to North Africa, or from Siberia to the Indopacific realm as well. But it is not necessarily true in reverse, in moving from the tropics to south temperate areas.

Take, for example, the family Anapidae, a group of tiny spiders, usually less than 2 millimeters long, that spin tiny orbwebs, usually less than a centimeter in diameter, in forest litter and moss. In the Nearctic region, there is only 1 genus, including only 1 species, restricted to California and Oregon (Platnick and Forster 1990). Three genera occur in the American tropics, 2 of which are endemic there, and the 3 genera together include at least 35 species (Platnick and Shadab 1978, 1979). But the south temperate fauna, in Chile and Argentina, is nowhere near as depauperate as in North America; it includes 6 endemic genera containing at least 15 species (Platnick and Forster 1989). The differences are even more dramatic in the Asian sector. In Asia itself, there are about 4 genera and 8 species, occurring from Nepal to Malaysia and Japan, and the tropical eastern Indopacific anapid fauna includes at least 2 genera and 5 species. But the south temperate fauna is by far the largest; New Caledonia has 3 endemic genera, with 8 species; New Zealand has another 3 endemic genera with 13 species; and Australia has another 10 endemic genera and 37 species (Platnick and Forster 1989). One of those Australian genera is known only from mountaintops in northern Queensland, and its 9 species, found on different mountains there, could debatably be considered tropical rather than south temperate. But the point is simply that, at least for these spiders, and I suspect for most nonvertebrate groups as well, the picture of a species-rich tropics and species-poor north and south temperate zones is inaccurate.

For a more accurate picture, consider the spider fauna of eastern New York and New England; it includes some 683 species (Kaston 1981) and seems close to being completely described. The spider fauna of the United States and Canada includes some 3,412 species, plus about 230 more that have not yet been described (Roth 1985). Now consider the spider fauna of New Zealand, which is comparable to the New York–New England area in number of square miles. The New Zealand spider fauna is much less thoroughly described, at this point, but estimates of 2,500 species were published some time ago (Forster and Forster 1973:67), and today it seems likely that there will turn out to be as many species in New Zealand as there are in the entire continental United States. Australia is smaller than the United States, even excluding Alaska, but its spider fauna has been estimated at more than 9,300 species (Raven 1988), compared with our 3,600. And spiders are hardly unique in this respect. There are probably more species of flowering plants, per unit area, in the Cape Province of South Africa than anywhere else—some 16,000 flowering plant species are found in South Africa, according to an old estimate (Acocks 1953). In short, there is a common conception of this planet and its biodiversity as being roughly egg-shaped, hefty in the middle and narrow at both ends, but that conception may not be all it is cracked up to be. That conception seems mostly to be an effect of the boreal and megafauna biases. There is probably a better model for the shape of the world's biodiversity; real species counts for all animal and plant groups taken together are probably shaped more like a pear than an egg—the preponderance of biodiversity is in the tropics *and* south temperate zones.

But answering the question of what parts of the world have the greatest biodiversity in general terms like "the tropics and the south temperate zones" is of little use in the current crisis, which requires much more specific answers. It might seem, at least to a general biologist, or to an ecological biogeographer, that more specific answers would be easy to obtain. If we sample two areas and find that one has more species or that it has a larger number of species that meet some minimal abundance criterion, is that area not automatically the one most worth preserving? From the vantage point of a comparative biologist, "it ain't necessarily so."

Consider instead three areas, including the two just mentioned. Suppose that they house, respectively, 5,000, 4,000, and 3,000 species (either in total or counting only those that occur abundantly enough to maintain population levels that might be judged adequate by some standard). Suppose also that for logistic reasons, we can hope to preserve only two of those areas. Should we automatically pick the first two, A and B? In fact, the number of species alone is an insufficient basis for a decision; we must consider what particular species are there, and what their total distributions may be, as well as how many of them there are.

Take a simple example. Suppose that the species lists for areas A and B show complete overlap; in other words, area A includes every species that occurs in area B, plus 1,000 more. Suppose also that area C has a very different fauna; 100 of its species are shared with areas A and B, but 2,900 of them are not. In that case, the combination of areas A and C obviously includes more biodiversity than A plus B. One can easily juggle the numbers so that area A, with the most species, is the worst choice for inclusion: in the case shown in figure 2.1, preserving the two least diverse areas might save 6,900 species, compared with 5,000 for the two most diverse areas alone.

In other words, it is not merely the number of species that counts but also the proportion of narrowly distributed or endemic species. That is an interesting result, because paying attention to areas of narrow endemism is the hallmark of modern vicariance biogeography, which basically asks three questions:

1. What are the smallest areas of the world that house endemic species—how many are there, and where are they? We have only fragments of an answer, but it looks as though there are many local areas of endemism, each defined by the overlap of two or more species ranges.
2. Given some number of areas of endemism, are the species occurring in them phylogenetically related to species elsewhere in some identifiable pattern?
3. If there are one or more general patterns of relationship among areas, do these patterns correlate with identifiable events in earth history?

On a very broad, global scale, these questions have been answered. There are general patterns of relationships; for terrestrial organisms these general patterns often span ocean basins and sometimes clearly reflect old seafloor-spreading events and subsequent continental drift. These global answers are again not particularly relevant to our current crisis, but other studies, done on a much finer scale, are entirely relevant. Consider, for example, the classic work done by the late Donn Rosen (1979) on the fishes of Central America. Rosen investigated two genera of freshwater poeciliids that are speciose in that part of the world, occupying 10 different areas that can be defined simply by the distributions of those species. He then analyzed the phylogenetic rela-

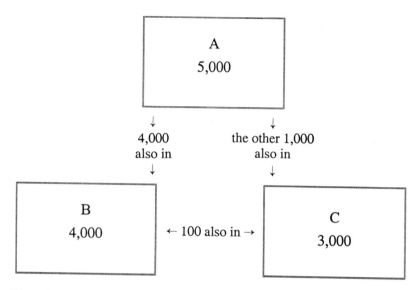

Figure 2.1 Hypothetical biodiversity of three allopatric areas, each containing both differing numbers of species and differing numbers of species shared with the other two areas.

tionships among the species of each genus. Subsequent comparisons showed that the genera agree in great detail on the interrelationships of the areas and that such detailed agreement could not have been expected by chance alone (Platnick 1981; Page 1989) but therefore presumably reflects a common cause—in this case, a common history, to which various groups have responded in differing, but cladistically congruent, ways. Similar degrees of congruence have been found, for example, among Australian eucalypts (Ladiges, Newnham, and Humphries 1989, and earlier papers cited therein) and birds (Cracraft 1986).

Areas of endemism vary in size, of course, but they can be alarmingly small. Also alarming is that they may tend to be smaller in south temperate areas than in the tropics. Cracraft and Prum (1988) looked at Neotropical birds, for example, and identified a pattern involving the interrelationships of several areas of endemism; those in Amazonia are quite large. Smaller Amazonian areas may well be recognizable on the basis of tropical American butterflies (Brown 1982) or other groups of nonvertebrates. Many Neotropical spider species seem to have narrow ranges, but this may be largely a reflection of grossly inadequate sampling efforts; certainly many other Neotropical spider species are quite widely distributed, occurring from the southern United States south to northern Argentina. In Chile, on the other hand, that is usually not the case (and species widespread in the rest of South America are usually absent). Some obvious areas of endemism in Chile are very small; forest habitats, for example, are extremely sparse in the northern half of that country. Some of those forests are fog-fed and occur only on a couple acres of mountain slope; luckily most of those relict forests are already protected in national parks and reserves. But other areas of endemism in Chile seem to reflect older events in earth history, rather than such obvious differences in present-day ecology. For example, there is a monophyletic group of eight species belonging to the ground spider genus *Echemoides;* these eight species are all allopatrically distributed in central Chile (Platnick 1983), even though there are no obvious ecological barriers separating them today. On the basis of the Chilean spider groups examined to date, there are at least twenty, and probably closer to forty, identifiable areas that house endemic taxa of different genera within central and southern Chile alone. By and large, the closest relatives of endemic Chilean spiders are found, not elsewhere in South America, but rather in New Zealand and Tasmania. Hence, "South America" is clearly not a natural biogeographic region (in the sense of a natural grouping of areas of endemism that are each other's closest relatives). Many Chilean areas of endemism are seriously threatened; forests have been burned in Chile, since the 1940s, at an appalling rate. On some occasions, I have had the unpleasant experience of trying to recollect, at some of my type localities, species I had taken just a few years earlier, only to find that what had been pristine *Nothofagus* forests were just piles of smoking embers.

In an ideal world, systematists and biogeographers would have enough resources, both financial and human, to sample adequately the entire world's biota, to describe all its species, to analyze all their relationships, to compare all their distributions, and at least to begin to understand both the historical and the present-day ecological factors that determine these distributions. Obviously, we do not live in that ideal world, and we cannot wait for such comprehensive knowledge before taking action in today's world. But it is important that we do not proceed only on the basis of an ecologically oriented approach to biodiversity that pays attention only to the numbers of species, or their relative abundances, in threatened areas and ignores the question of which threatened areas with high numbers of species are also areas of high endemism. And it is equally important that we stop portraying the problem as one that exists only in the tropics. We must admit that threatened areas in southern South America, in southern Africa, New Caledonia, and even Australia and New Zealand, as well as some in the north temperate zone, may imperil significantly large and (more importantly) unique components of biodiversity. We must face the possibility that at least some of these temperate areas may show a greater degree of local endemism than many tropical sites. In a triage situation, we simply cannot afford to sustain the boreal and megafauna biases that have in the past distorted our picture of biodiversity on a global scale.

REFERENCES

Acocks, J. P. H. 1953. Veld types of South Africa. *Memoirs of Botanical Survey of South Africa* 28:1–192.

Brown, K. S., Jr. 1982. Paleoecology and regional patterns of evolution in Neotropical forest butterflies. In G. T. Prance, ed., *Biological Diversification in the Tropics*, pp. 255–308. New York: Columbia University Press.

Cracraft, J. 1986. Origin and evolution of continental biotas: Speciation and historical congruence within the Australian avifauna. *Evolution* 40:977–996.

Cracraft, J. and R. O. Prum. 1988. Patterns and processes of diversification: Speciation and historical congruence in some Neotropical birds. *Evolution* 42:603–620.

Forster, R. R. and L. M. Forster. 1973. *New Zealand Spiders: An Introduction.* Auckland: Collins.

Humphries, C. J. and L. R. Parenti. 1986. *Cladistic Biogeography.* Oxford: Clarendon Press.

Kaston, B. J. 1981. Spiders of Connecticut, rev. ed. *Bulletin of the State Geological and Natural History Survey of Connecticut* 70:1–1020.

Ladiges, P., M. R. Newnham, and C. J. Humphries. 1989. Systematics and biogeography of the Australian "green ash" eucalypts (Monocalyptus). *Cladistics* 5:345–364.

MacArthur, R. H. 1972. *Geographical Ecology: Patterns in the Distribution of Species.* New York: Harper & Row.

Nelson, G. J. 1978. From Condolle to Croizat: Comments on the history of biogeography. *Journal of the History of Biology* 11:269–305.

Nelson, G. and N. Platnick. 1981. *Systematics and Biogeography: Cladistics and Vicariance.* New York: Columbia University Press.

Page, R. D. M. 1989. Comments on component-compatibility in historical biogeography. *Cladistics* 5:167–182.

Platnick, N. I. 1981. Widespread taxa and biogeographic congruence. In V. A. Funk and D. R. Brooks, eds., *Advances in Cladistics: Proceedings of the First Meeting of the Willi Hennig Society,* pp. 223–227. Bronx: New York Botanical Garden.

Platnick, N. I. 1983. A review of the *chilensis* group of the spider genus *Echemoides* (Araneae, Gnaphosidae). *American Museum Novitates* 2760:1–18.

Platnick, N. I. and R. R. Forster. 1989. A revision of the temperate South American and Australasian spiders of the family Anapidae (Araneae, Araneoidea). *Bulletin of the American Museum of Natural History* 190:1–139.

Platnick, N. I. and R. R. Forster. 1990. On the spider family Anapidae (Araneae, Araneoidea) in the United States. *Journal of the New York Entomological Society,* 98:108–112.

Platnick, N. I. and M. U. Shadab. 1978. A review of the spider genus *Anapis* (Araneae, Anapidae), with a dual cladistic analysis. *American Museum Novitates* 2663:1–23.

Platnick, N. I. and M. U. Shadab. 1979. A review of the spider genera *Anapisona* and *Pseudanapis* (Araneae, Anapidae). *American Museum Novitates* 2672:1–20.

Raven, R. J. 1988. The current status of Australian spider systematics. *Miscellaneous Publications of the Australian Entomological Society,* 5:37–47.

Rosen, D. E. 1979. Fishes from the uplands and intermontane basins of Guatemala: Revisionary studies and comparative geography. *Bulletin of the American Museum of Natural History* 162:267–376.

Roth, V. D. 1985. *Spider Genera of North America.* Gainesville, Fla.: American Arachnological Society.

3 : Systematic Versus Ecological Diversity: The Example of the Malagasy Primates

Ian Tattersall

Diversity is a word so readily used by biologists that it rarely seems necessary to stop to consider the complexities underlying the concept it denotes. That the natural world is a very diverse place indeed seems intuitively obvious, but it is diverse in a variety of ways and on a variety of scales, so that our usages of the concept may not in fact be comparable from one context to another. It is also true that while one usually has no difficulty in recognizing diversity when one sees it, in most cases one is hard pressed to characterize it with adequate precision, and to me it seems clear that it is within this specific context that museums are uniquely fitted to contribute to our understanding.

I shall return to this theme later, but I should begin by addressing the more specific concerns of this session, which center around the value of distinguishing between different kinds of diversity, specifically between its phylogenetic and ecological forms. Actually, it would seem necessary also to distinguish between different kinds of diversity within each of these categories. Thus diversity in its systematic sense cannot be characterized simply as the sheer abundance of taxa contained within a monophyletic group, because a plethora of taxa at low taxonomic levels can have implications for past and/or continuing processes that are very different from high diversity measured at, say, the ordinal level or above. Similarly, while we may use species counts as a measure of diversity either within a particular higher taxon or in a particular geographic region, the exact mechanisms underlying the two varieties of diversity thus measured may be very different, even though both ultimately result from the interplay of ecological and historical factors. Nonetheless, although ecology and history play conceptually different roles in producing diversity, it is undeniable that in producing past and present biotas these two factors have been closely intertwined, perhaps even inextricably.

In one clearly definable sense the history of life is the sum total of the history of all those taxa that have ever existed, and it can be indeed be summarized in statements of the relationships among those taxa. But whereas this may provide the skeleton of the history of life, its living, breathing body is clothed in the interactions among those taxa in existence at any one time and place and in the interactions between these players and the abiotic world in which they lived or live. A complete understanding of the story of life on this planet requires a synthesis of all these factors, and a synthesis requires that all be considered within the same framework. But how can we synthesize systematic studies of the diversities of monophyletic taxa over time and space with ecological studies of biotas, agglomerations of species coexisting within specific environments? Even if we regard historical factors as ultimately representing the outcome of all those ecological processes that have operated in the past—a formulation that would certainly not be acceptable to everyone—can we achieve a comprehensive view of biodiversity into which our studies as ecologists or systematists will fit?

On the face of it, this might not seem too difficult; after all, systematists and ecologists share a fundamental aspect of their terminology: both describe the subjects of their study in terms of species, whether as parts of communities or as components of monophyletic groups. On the other hand, Eldredge (1986 and this volume) has argued eloquently that ecological and systematic systems in nature differ in essence because they are based upon fundamentally different aspects of the activity of organisms: the economic and the reproductive, respectively. The implication of this is that species mean very different things to ecologists and systematists, and in an operational sense this is undoubtedly true. But is it true in a more profound sense, one that would suggest that synthesis of these elements in describing biodiversity might be more difficult to achieve than we would like to think? Clearly, a link between the two systems exists: individuals are both economic interactors and reproductive entities, even though the factors that affect them in these two roles are very different. But the individual is hardly a satisfactory basis for either ecological or systematic studies, and it is thus important to determine whether the two systems intersect at more inclusive levels. Expressed this way, the question becomes a more empirical one: what concordance can we find in nature between components of the two systems that might suggest that common processes underlie them?

Madagascar: A Microcosm

Empirical questions of this kind require consultation of nature, rather than of navels. One way of approaching the problem in a very general way is to compare the distributions of related taxa over a given set of ecological zones, to determine whether taxic and ecological boundaries are in any way coincident. Madagascar, a thousand-mile-long island lying some 300 miles off the southeastern African coast (figure 3.1), and almost entirely within the southern tropical zone, provides an excellent setting for such a comparison. But this island, with its large size and varied topography, does not only offer us a remarkable array of habitats occupied by a highly endemic fauna. For as one of the world's unique and most highly endangered environments, Madagascar,

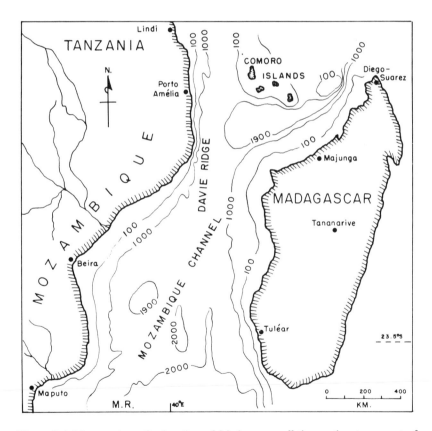

Figure 3.1 Map to show the location of Madagascar off the southeastern coast of Africa (Bathymetry in meters)

in its current plight, emphasizes the underlying theme of this conference: that the world's biodiversity, however we define it, is under severe siege and that time left to us in which to reverse a disastrous process of destruction is critically limited.

Isolated from the neighboring continent in a process that began some 165 million years ago and that was completed before the dawn of the Age of Mammals, Madagascar has a mammalian fauna that is depauperate in major groups: among its truly indigenous land mammals only Primates, Insectivora, Carnivora, Rodentia, and Chiroptera are represented today, although until not long ago a pygmy hippopotamus seems to have been relatively common. And while the sole indigenous carnivore and insectivore families, Viverridae and Tenrecidae, are both quite rich in species, the only Malagasy mammal order that is strikingly diverse is Primates. The extant primate fauna of Madagascar, collectively known informally as the lemurs, contains 5 totally endemic families, embracing some 30 species classified in some 13 or 14 genera, and if we add in readily identifiable subspecies, we reach a total of about 50 distinguishable taxonomic units. Moreover, until only about a thousand years ago, not long after the arrival of humans on the island, there existed at least an additional 8 genera and 15 species, some of which belonged to families that are now totally extinct. Virtually all of these extinct lemurs were larger in body size than those still surviving, and they were almost certainly the victims of human interference (e.g., Tattersall 1982; in press). Today we are thus witnessing not simply a Malagasy primate fauna that is on the brink of disappearance, but also one that is already partway through an ongoing process of anthropogenic extinction.

By any measure the diversity of Madagascar's primate fauna is thus remarkable, and it is a diversity that must surely be traceable to a combination of historical and ecological factors. The principal historical factor is time: although it appears unlikely that all the lemur families are descended from a single Malagasy ancestor (Tattersall 1973, 1982; Schwartz and Tattersall 1985), all have certainly been established in the island for a long time—conceivably since as long ago as the late Cretaceous, and certainly since the Eocene (Tattersall 1982). A minimum of 50 million years has thus been available for the diversification of the lemur fauna. Unfortunately, however, the lack of a pre-Holocene mammalian fossil record in Madagascar leaves us in total ignorance of the longevity of the various contemporary lemur taxa, and our knowledge of past climate and vegetational change in the island is similarly limited, although the efforts of David Burney and his colleagues (e.g., Burney 1987) are beginning to rectify this, at least as far as the Holocene and latest Pleistocene are concerned.

This dearth of knowledge clearly bedevils any attempt to tie in any but the most recent episodes of taxic diversification in Madagascar with potential ecological factors. But it nonetheless appears intuitively reasonable to assume that Madagascar's lemurs could not have diversified as they have done in the absence of the remarkable variety of habitats that characterizes the island today and that presumably did so also in the past.

Madagascar: The Environment

Let us, then, look briefly at the spectrum of environments found in Madagascar today. Various studies at different times (e.g., Humbert 1955; Humbert and Cours Darne 1965; Koechlin, Guillaumet, and Morat 1974) have divided Madagascar into detailed series of bioclimatic or phytoecological zones; Humbert's (1955) broad division into phytogeographic zones, elaborating an earlier schema by Perrier de la Bathie (1921), seems most appropriate for present purposes and is shown in simplified form in figure 3.2. Note that the central part of the island, a rugged plateau with an average elevation of around 1,500 m, has been largely denuded of forest since the arrival of humans in Madagascar (though to what extent the area was clothed with forest at that time is still debated: see Burney 1987) and need not be considered here since the surviving lemurs are fairly closely tied to the forested environments that form a "wreath" around the island's periphery. The aboriginal vegetation in the peripheral areas of Madagascar has also been dreadfully ravaged, and the ranges, or at least the surface areas, occupied by virtually all the surviving lemurs have been drastically reduced over the past millennium or so.

Broadly, the eastern side of Madagascar, which consists of the eastern edge of the plateau and a rugged escarpment that drops toward a narrow coastal plain and which catches the prevailing easterly winds, is an area of high and consistent rainfall. The Western Region, in which the plateau slopes down more gently into a pair of broad sedimentary basins, experiences considerably lower rainfall, seasonally distributed. Within the Eastern and Western Regions, however, Humbert recognized a series of distinct Domains, each with its own vegetational characteristics. Distinctions between the various Domains appear to be controlled by climate; variations within each Domain, which may be considerable, and which were at least partially reflected by Humbert in a further subdivision into Sectors (not shown in Figure 3.2) are related principally to edaphic factors.

In the west, the semiarid Southern Domain is characterized by thickets or forests of highly endemic spiny, bushy, and xerophytic vegetation, dominated by species of Euphorbiaceae and Didiereaceae. Deciduous canopy forests, with relatively open floors, also occur locally, mostly along river banks. In the Western Domain, three types of forest occur: predominantly river-valley deciduous canopy forests similar to those found in the Southern Domain and

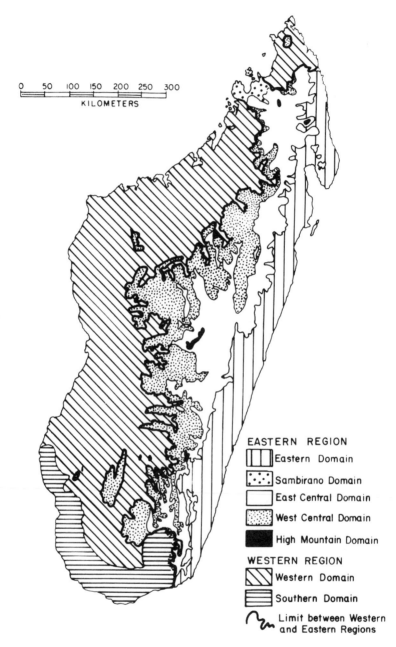

0   50   100 150 200 250 300
KILOMETERS

EASTERN REGION

⫼ Eastern Domain

⫶ Sambirano Domain

□ East Central Domain

▨ West Central Domain

■ High Mountain Domain

WESTERN REGION

▧ Western Domain

▤ Southern Domain

〰 Limit between Western and Eastern Regions

Figure 3.2 Phytogeographic divisions of Madagascar, as established by Humbert (1955)

generally dominated by tamarind trees (whose antiquity in Madagascar is a matter for investigation); on drier soils, more xerophytic forests with dense undergrowth that lack a well-defined canopy and dominant species; and on the driest soils of all, species-poor thickets of brush-and-scrub. The frequency of the moister vegetation types tends to decline from north to south.

The Eastern Domain is typified by dense stratified evergreen rain forest with a relatively open floor and abundant epiphytes. Toward the south, especially, this forest has become widely replaced by dense heliophilous formations in the wake of human disturbance. The forest of the Sambirano Domain on the northwest coast is in many ways similar to that of the Eastern Domain, but a pronounced dry season of some three to four months' duration gives it a somewhat drier aspect. The northern tip of Madagascar is extremely arid, while south to the Sambirano the vegetation resembles that of the Western Domain, although the Mt d'Ambre, in the center of this area, is clothed in evergreen rain forest.

The Malagasy Primates

I have briefly outlined the extent of the taxic diversity of the lemurs above, and I have noted that the absence of a fossil record makes it difficult to calibrate the diversification of this fauna in time. Further, I have pointed out that our knowledge of past environments in Madagascar is negligible. The combination of these factors suggests that any attempt to correlate ecological processes with processes of taxic diversification in Madagascar must be confined to the most recent episodes of diversification, i.e., those that are most likely to have taken place in environments comparable to those currently existing in Madagascar. It is probable that much of the differentiation we see among the lemurs at the species level and below dates from the Pleistocene, during which it is inconceivable that the island escaped the marked climatic swings that affected environments worldwide. But while environmental shifts in Madagascar during this epoch certainly involved the expansion and contraction of habitats that we can recognize today, they are unlikely to have involved a radically different set of habitats. Which of Madagascar's primates, then, might it be potentially helpful to examine in this context?

Of the 13 or 14 extant lemur genera, only 3 are notably diverse at the species level and below. These are *Lepilemur, Lemur,* and *Propithecus.* The first of these, while by primate standards unquestionably rich in species, is currently something of a taxonomic basket case (though in the course of revision: Tattersall and Schwartz, in preparation), and I am reluctant for this reason to examine it here. The genus *Lemur,* however, contains 6 species (see figure 3.3), all of which are monotypic in current classifications except for *L.*

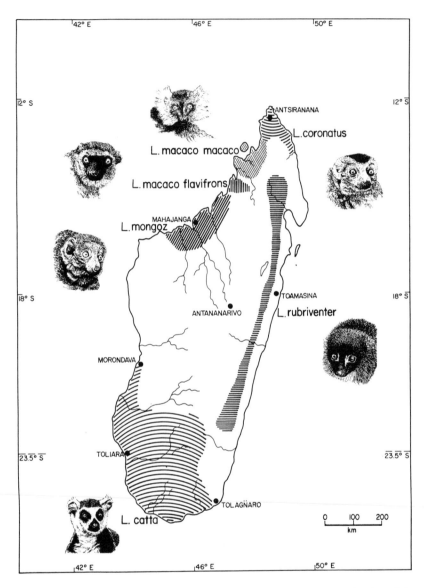

Figure 3.3 Map of Madagascar to show the approximate areas of distribution of the species of *Lemur*, with the exception of *Lemur fulvus*. Distributions are not necessarily continuous within the areas indicated.

macaco, which boasts 2 subspecies, and *L. fulvus* (see fig. 3.3), which contains 6. Both *Lemur* species and *Lemur fulvus* subspecies (see figure 3.4) are distributed around most of the periphery of Madagascar. Similarly, the ubiquitous genus *Propithecus* (fig. 3.5) contains 3 species, 2 of which each contain 4 subspecies.

In all cases these populations, both at the species level and below, are readily differentiable on the basis of characters of color and external morphology. At present, however, it remains difficult to specify the geometry of the relationships among the species and subspecies involved, so the construction of area cladograms comparing patterns of relationship with patterns of distribution is impractical. Further, our knowledge of precise distributions, both ecological and geographic, is far from perfect, and it is not even certain that we have at this point appropriately identified in all cases the levels of differentiation between closely related populations. In trying to find common patterns underlying systematic versus ecological diversity in the lemur fauna we are for the moment left, therefore, with the rather blunt instrument of broad comparisons of the ecological with the geographic distributions of differentiated populations.

Lemur Distributions and Lemur Ecologies

A comparison of the geographical distributions shown in figures 3.3–3.5 with Humbert's phytogeographic zones shown in figure 3.2 shows immediately that there is no necessary concordance between systematic and ecological boundaries. Of the three *Propithecus* species, *P. diadema* is largely limited to Humbert's damp Eastern Region, but *P. d. perrieri* represents an outlier in one of the most arid parts of the island. And while *P. verreauxi* is found only in the Western Region, the subspecies *P. v. verreauxi* straddles the Western and Southern Domains. This subspecies is found throughout the broad spectrum of environments offered by the Southern Domain, which as I have noted contains both highly xerophytic semiarid formations and deciduous riparian forests, and it continues northward into the moister regions of the Western Domain. *Propithecus tattersalli* exists as a tiny isolate within the northern extension of the Western Region, but even within its highly localized distribution appears to be found in forests of greatly varying aspects (D. Meyers, personal communication).

If we consider *Lemur fulvus* in its various guises simultaneously with the other species in its genus, we find representatives of *Lemur* in virtually all forested regions of Madagascar. *Lemur coronatus* is restricted to the northern extension of the Western Domain, where it is the only primate to be found in the extremely arid Cap d'Ambre, at Madagascar's northern tip. It also occurs, however, in the rain forests of the Mt d'Ambre and in all the intermediate

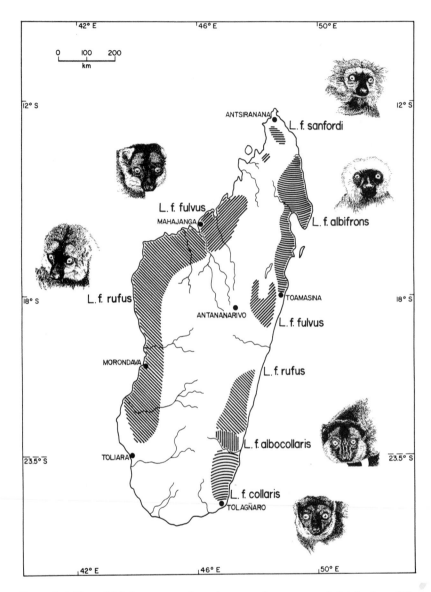

Figure 3.4 Map of Madagascar to show the approximate areas of distribution of the subspecies of *Lemur fulvus*. Distributions are not necessarily continuous within the areas indicated.

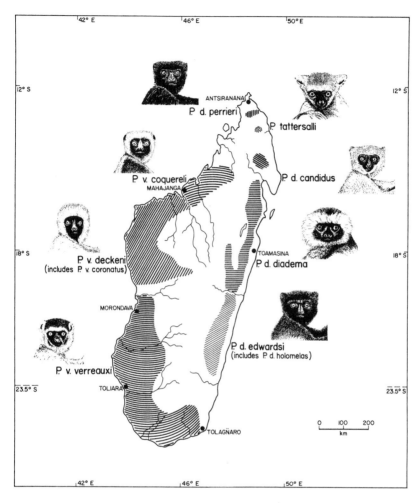

Figure 3.5 Map of Madagascar to show the approximate areas of distribution of the species and subspecies of *Propithecus*. Distributions are not necessarily continuous within the areas indicated.

environments of the region. Farther to the south, the parapatric subspecies of *Lemur macaco* are more or less restricted to the Sambirano Domain, but they flourish in degraded secondary environments along the coast as well as in the moister forests of the interior. *Lemur mongoz* is found in the relatively dry Western Domain to the south of the Sambirano but also occurs on the Comoro Islands, where it is (or was) found in great abundance not only in drier secondary formations but also in the montane rainforest of the island of Anjouan (Ndzouani) in the Comoro group. Continuing south, *Lemur catta* is found not only in all the environments offered by the Southern Domain (although, unlike *Propithecus verreauxi,* it may not occur in habitats consisting uniquely of highly xerophytic vegetation), but its area of distribution continues northward well into the Western Domain.

The subspecies of *Lemur fulvus* present us with the most remarkable picture of all. Two of these, *L. f. rufus* and *L. f. fulvus,* are found in forests of both the Western and the Eastern Regions (which must thus in the past have been connected at least intermittently across the center of the island, possibly by tracts of riverine vegetation if and/or when the center was not continuously forested). Both thus occur in seasonal deciduous forests and in evergreen rain forest. *Lemur f. fulvus* is also found, moreover, in the Sambirano Domain and flourishes additionally (under the guise of *L. f. mayottensis*) in the degraded secondary habitats of Mayotte, in the Comoro archipelago. *Lemur fulvus sanfordi* occurs not only in the rainforests of the Montagne d'Ambre in the northern extension of the Western Domain but also in the dry forests of the adjacent lowlands.

Detailed studies currently underway will, it is hoped, eventually allow cross comparison of patterns of historical differentiation in these lemurs with their geographical distributions, but for the moment, even this rather simpleminded comparison of the latter with habitat conditions makes one important point with, I think, some force. This point is that habitat preference—ecology—cannot be viewed as a property of any of these taxonomic groups, even at the low level of differentiation represented by the subspecies or the local population. To put it the other way around, we cannot use aspects of ecology to help us define or recognize lemur species or subspecies. I might add that my experience in studying lemur behavior suggests that the same holds for most aspects of behavior, too. But does this in fact reinforce the notion of the separateness of systematic and ecological systems in nature: the notion that systematic and ecological forms of diversity not only are different but also are largely or entirely unrelated, except perhaps to the extent that individuals are players in both games?

I doubt it. For fundamentally, both ecologists and systematists use the same essential unit of study, and it is not the species. The most the ecologist actually sees in the field is the local population, acting within the local community. Similarly, while the systematist may claim to be studying species, this larger unit is for practical purposes inferential rather than observable. Whether as systematists we subscribe to an exclusive notion of the species, such as the biological species concept, or to an inclusive one, such as the recognition concept, we cannot often, if ever, know absolutely whether the aggregates of individuals we study conform in practice to the requirements of either. While sympatric occurrences can usually tell us whether a pair of populations is reproductively isolated—although hybrid zones can still yield equivocal situations—we cannot directly read the limits of reproductive exclusivity from the materials of systematic study. Similarly, specific mate recognition systems may be hard indeed for us to identify and will not often be reflected in the specimens on which most systematic study is based.

In practice, then, species are essentially abstractions to both ecologists and to systematists: something that one never sees, they are rather the result of a process of extrapolation. In this light, it is hardly surprising that the economic and genetic roles of species are not viewed by all as coterminous. But local populations, in contrast, are both reproductive and economic units (though, perhaps surprisingly, their subcomponents may not be; *Propithecus verreauxi* troops, for example, appear not to be reproductive units in a strict sense [cf Richard 1978]), never more so than among mammals as social as the primates. At the origin of biodiversity we have local populations simultaneously acting in both economic and reproductive roles, and to the extent that these roles are the properties of the same entities, they cannot be entirely independent. It is at this level, then, that of the local population, that the two systems, the economic and the reproductive, intersect; they are joined by simultaneity in time and place, and hence, are connected by history.

All the lemur taxa I have mentioned are descendants of other taxa. These taxa must have been distributed over a variety of ecologies, just as their descendants are. The local populations that gave rise to those descendants were almost certainly fractions of larger populations, with adjacent portions of which they interacted reproductively, at least for most of their histories. It was probably largely historical accident that determined which of those populations survived to produce descendants: historical accidents involving, perhaps, local environmental change or the relative efficiency of neighboring competitors. These historical factors may justifiably be defined as ecological: extrinsic factors, as opposed to the intrinsic factors that promote reproductive cohesion. There is, thus, a common element in both systematic and ecological systems: the evolutionary history of any group sums out as the history of more or less successfully reproducing local populations interacting economically with the environment, biotic and abiotic, around them. Conversely, ecological communities are necessarily composed of entities supplied by the lines of descent of the organisms involved. At higher, more abstract levels we lose this common element, but at the level of the local population the twin threads of the history of biotic diversification do appear to intertwine.

The Role of Museums in Promoting Understanding of Biodiversity

Natural history museums, for obvious reasons, have traditionally devoted themselves to the business of systematics. After all, you cannot collect ecologies. Their importance as resources for systematists is primordial and acknowledged as such, making it unnecessary to elaborate this point here. But

in a world where both ecological communities and large numbers of systematic groups are under threat everywhere, museums will in some areas at least
find it increasingly difficult, or often impossible, to continue collecting the
kinds of material of which they have built up their collections over the past
couple of centuries. And indeed, no responsible museum professional would
wish to add to the stresses upon populations already on the brink of disappearance. Thus a change of emphasis seems not only desirable but also mandated.
Obviously, a first priority must be on the adequate preservation and curation
of material already held, so that these irreplaceable resources can continue to
be the foundation of researches into the systematic aspects of biodiversity. But
beyond this, there are other ways to enhance the values of existing collections
as scientific resources. For museums should not simply be concerned with
collecting objects; they should be actively promoting the acquisition of information about the natural world of the kind that will enhance the scientific
value of their collections.

This is true because, as what I have said above suggests, museum specimens are simply the core of a much larger context, which embraces not only
systematics but ecology as well. We can all agree that almost any specimen
without that most basic bit of associated information, locality, is next to useless for systematic purposes, and it is equally clear that the utility of any specimen increases with the amount of information that accompanies it. Further,
it is still possible today to collect information that will augment the value of
existing specimens, even if collection of new specimens themselves is impossible or inadvisable. If ecological and systematic systems in nature are indeed
interdependent, then knowledge of one must at some level shed light on the
other. It follows that to maximize the possibilities inherent in existing collections, museums should make every effort to augment the information available
on the contexts—in the broadest sense: geographical, ecological, social—in
which the individuals now preserved in their collections lived their lives.

One final point, specifically about conservation. Except in the limited context of captive conservation (which is invariably conducted today with one eye
on reintroduction of populations into their natural environments once the latter
have been stabilized), it is impossible to conserve species selectively. Habitats—ecologies—have to be conserved as wholes if the organisms that make
them up are to survive. And to conserve, we must know what it is that we are
conserving; we must be able to characterize the diversity we wish to protect.
Ecologies consist of the organisms we identify in systematic terms and of
the interactions among them. Economic interactions are the business of ecologists, but in the specification of the interactors, museums are uniquely
equipped to contribute. Museum systematists, and the collections they study,
thus have an important role to play in describing exactly what it is that we
wish to conserve and in identifying those regions of the world that it is of

particular importance to protect. Others will make the points that many of the most threatened environments are in developing countries that currently lack the expertise and resources to play this role themselves and that natural history museums in the developed countries are uniquely equipped to help third-world governments appreciate the importance of systematics and to train local systematists in those areas where local expertise is lacking. I would simply observe here that natural history museums, while priding themselves on being libraries of the diversity of the living world, are in danger of simply becoming history museums, containing records of a diversity that has vanished. Clearly, it is our responsibility to do everything possible to prevent this from happening.

REFERENCES

Burney, D. A. 1987. Presettlement vegetation changes at Lake Tritrivakely, Madagascar. *Palaeoecology of Africa and the Surrounding Islands* 18:357–381.
Eldredge, N. 1986. Information, economics and evolution. *Annual Review of Ecology Systems* 17:351–369.
Humbert, H. 1955. Les territoires phytogéographiques de Madagascar, leur cartographie. *Annales de Biologie* (3d sér.) 31:195–204.
Humbert, H. and G. Cours Darne. 1965. Notice de la carte Madagascar, Carte International de Tapis Végétal. *Travaux Section Scientifique Institut Français Pondichéry*. Hors sér. 6:1–162.
Koechlin, J., J.-L. Guillaumet, and P. Morat. 1974. *Flore et Végétation de Madagascar*. Vaduz, Liechtenstein: J. Cramer.
Perrier de la Bathie, H. 1921. La végétation Malgache. *Annales du Musée Colonial de Marseille* (3d sér.) 9:1–268.
Richard, A. 1978. *Behavioral Variation: Case Study of a Malagasy Lemur*. Lewisburg, Pa: Bucknell University Press.
Schwartz, J. H. and I. Tattersall. 1985. Evolutionary relationships of living lemurs and lorises (Mammalia, Primates) and their potential affinities with European Eocene Adapidae. *Anthropological Papers of the American Museum of Natural History* 60(1):1–100.
Tattersall, I. 1973. Subfossil lemuroids and the "adaptive radiation" of the Malagasy lemurs. *Transactions of the New York Academy of Sciences* 35:314–324.
Tattersall, I. 1982. *The Primates of Madagascar*. New York: Columbia University Press.
Tattersall, I. In press. Patterns of origin and extinction in the vertebrate fauna of Madagascar. In M. Sanges and P. Y. Sondaar, eds., *Early Man in Island Environments*.

4 : Spilling Over the Competitive Limits to Species Coexistence

George Stevens

If you were to get in your car and drive from Philadelphia, Pennsylvania, to Filadelfia, Costa Rica, you would witness an enormous increase in the variety of life. The great species richness of tropical lands as compared with their temperate and polar counterparts is hard to exaggerate. An intense day of bird watching in the woods might yield 35 species in upstate New York but more than 300 in Costa Rica (Van Velzen and Van Velzen 1988). The immediate questions to an ecologist are: How can so many species coexist in tropical habitats? Why don't more species co-occur in temperate and polar ones? What are the limits to species richness?

The latitudinal gradient in species richness can also be illustrated by comparing the number of species in a given taxon at different latitudes (figures 4.1–4.3). Trees, birds, mammals, fish, mollusks, insects—a wide variety of higher taxa show a pattern of increasing species richness with decreasing latitude. This pattern breaks down when lower order groupings like genera are compared, but even so these findings still prompt the question: Why are so many more species of widely distributed taxa found in the tropics as compared with temperate and polar areas? What are the limits to species richness?

While it would seem the two biologists have asked the same question, they have actually asked one question about mechanics and another about history. The distinction between mechanics (literally the action of forces on a body) and history (the long-term outcome of those interacting forces) separates the perspectives of evolutionary ecologists and macroevolutionists and also happens to effectively segregate the varied explanations for the latitudinal gradient in species richness (reviewed by Pianka 1966; Brown and Gibson 1983). On the one hand we have the mechanistic explanations: different niche widths, differential rates of predation, and differences in overall productivity. On the other hand are the historical explanations: differential recovery from past glaciations, differing rates of extinction and speciation, and differences in each latitude's long-term ecological stability.

The debate over the causes of the latitudinal gradient in species richness has been the longest standing unresolved puzzle of community ecology. More than 100 years of speculation has produced a dozen or so viable nonmutually exclusive hypotheses. No one explanation has come to the forefront because each hypothesis grades imperceptibly into its neighbor on the mechanical-to-historical spectrum. The interconnectedness of the hypotheses arises from the thread of time that connects each historical explanation to a mechanical one. This is because, given sufficient time, mechanical processes acting on individuals become historical patterns reflected by species and higher taxa.

It would be easy to overemphasize the similarities between the latitudinal gradient debate and the choosing up of sides in the building turmoil over taxonomic versus ecological limits to species richness (the latter outlined by Eldredge and Cracraft 1980). Nevertheless, it is striking how effortlessly the opponents in either dispute can be sorted into the "historian" and "mechanic" categories. The similarities suggest that we are headed for many more years of unresolved speculation, this time generated by the biodiversity crisis. Before entering the fray, maybe we should stop and ask whether the ground rules of the debate need revision. Why echo the discussions of our intellectual grandparents?

If we ignore the biology for a moment and focus just on the kinds of hypotheses that have been proposed to account for the latitudinal gradient in species richness, it is clear that "time" in all its aspects has been pretty well worked over by those who originally structured the debate. We have the anomalies of the present (harsh environments) and the anomalies of the past (glaciation) being used to explain why extratropical latitudes have fewer species than the tropics. Competition now and narrow niche breadth (a result of competition in the past) have both been used to explain why tropical habitats contain so many species. It seems that every mechanical hypothesis has a matching historical hypothesis. If treated just as a puzzle it should be clear that neither side won the argument, because their suggested solutions differed only by their choice of timescale. The arguments deteriorate into contests to resolve which takes priority: process or pattern.

Furthermore, the long duration of the debate and the creative thinkers who

Breeding Bird Species

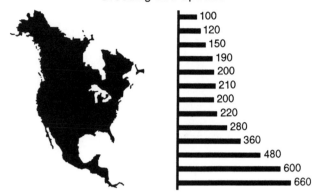

Tree Species

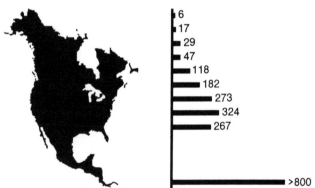

Mammal Species

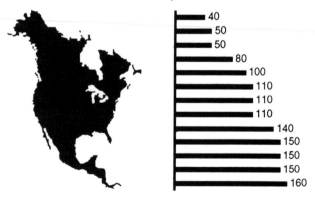

have applied themselves to the problem suggest it is extremely unlikely that a simple solution using the present format of the debate will ever be found. For example, consider competition as a limit to species richness. There can be no denying that competition is important in some systems (reviewed by Connell 1983). Even though the mechanistic aspects of competition grade imperceptibly into the historical consequences of continuous competition, competition is still an important force. That the explanations generated thus far for the latitudinal gradient in species richness are tangled and interconnected does not mean that they are wrong. The message for both the discussion of latitudinal gradients and the issues of taxonomic versus ecological limits to biodiversity is that to make progress we must not concentrate on time but instead jointly reformulate the debate.

In the discussion of issues so inextricably linked together, progress comes when the framework of the debate is altered rather than by attempts to test long-standing hypotheses. So far time has been mined for insights much more than space. Both time and space are components of any quantification of species richness. The number of birds in a woodlot in a given year or the number of species in a given taxon in the Cretaceous both have a spatial component— the boundaries of the woodlot in the former, the breadth of the sampled habitats open to the taxon in the latter.

Biogeographers study how space determines species richness in a given locale through the study of overlapping geographical ranges. In contrast, a field biologist selects a study site and looks for the interactions on that piece of ground that reflect the organizational rules of the community. Although some ecologists have tried to forge union between ecology and biogeography (see especially Brown 1988), biogeographical concepts are not well integrated into local level studies. Many ecologists used to consider the geographical range of a species as nothing more than the summation of all points where the species is able to hold its own.

All of this changed rather abruptly with the realization that habitats can be characterized as "sources" or "sinks" (Pulliam 1988) for different species. With these changes the boundaries of the study site are being broken down.

FACING PAGE:
Top: Figure 4.1 Number of breeding bird species found at different latitudes of North and Central America (data in Cook 1969).
Middle: Figure 4.2 Number of tree species found at different latitudes in North and Central America (North American data from Brockman 1968; Central American point is a conservative estimate derived from several sources: Allen 1956; Holdridge and Poveda 1975; and Croat 1978).
Bottom: Figure 4.3 Number of mammal species found at different latitudes of North and Central America (data in Simpson 1964).

No longer can we imagine that the point sample is all we need to understand the species richness of the point.

In retrospect, an obvious illustration of this fact is found on any mountain tall enough to include a treeline habitat (Arno 1984). Tree line is the upper elevational limit to tree distribution. Higher elevations are too harsh for trees to tolerate. Just below treeline is a less visible boundary that is the upper limit of where individual trees are regularly able to produce viable offspring. The elevational distance between the tree-reproductive limit and treeline varies depending on the species but has its origin in the ability of many organisms to forgo reproduction when resources are limiting. The difference between the requirements for continued life versus those needed for life and reproduction account for the narrow band of physiologically alive but genetically dead at treeline. These spillover trees are the living dead (D. H. Janzen, personal communication). They arrived in their predicament through migration into the area (as seeds) from lower elevations where their parents were reproductively active. Said differently, the habitat of the living dead is a sink. The source is somewhere at lower elevation.

If one was to study only the living dead, it would be very difficult to figure out how their population is being maintained. They are losing in competition, but still they continue to be represented in the area and take up resources. They never reproduce, but even so their population contains many juveniles. The only conclusion is that the number of living dead in a sink habitat is decoupled from any processes occurring in that habitat.

No matter what the ultimate cause of a geographical boundary or a habitat boundary to the distribution of a species, there will always be individuals that disperse out of the source and into the sink. Treeline is just the most visible manifestation of a phenomenon that occurs whenever a species reaches its distributional limits. We should expect the living dead to form a ring around all geographical ranges.

The importance of this recent realization for questions of species richness is seen when the geographical range of different species is plotted as a function of latitude. The average size of the geographical range of species decreases dramatically with decreasing latitude (figures 4.4–4.8). With the many small geographical ranges in low-latitude areas the potential for numerous overlapping sinks is greater than for higher latitudes. This comes about because with the decrease in geographical range size, the relative proportion of the range that is composed of sink habitats increases. This proportion changes because the ratio of circumference to surface area is negatively correlated with the size of the geographical range. This implies that the number of living dead in local species counts should increase with decreasing latitude.

The phenomenon of decreasing latitudinal breadth of species at decreasing latitudes is called "Rapoport's rule" (Stevens 1989) in honor of Eduardo H.

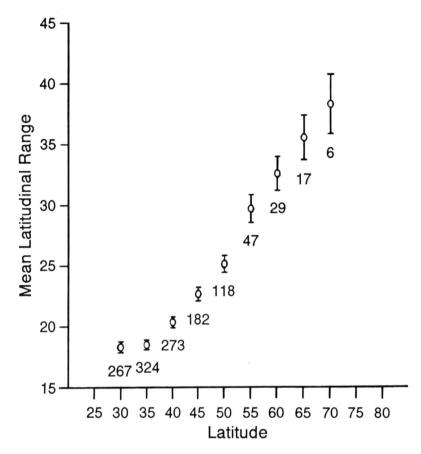

Figure 4.4 Mean latitudinal extent of North American trees found in different latitudes. In this and in figures 4.5–4.8 the width of the latitudinal bands is 5 degrees. Occupation of a band was determined by dividing the geographical range of each species into strips of 5 latitudinal degrees and determining overlap at any longitude. Error bars are one standard error of the mean. Numbers below the bars are sample sizes.

Rapoport, the Argentinian biogeographer whose work provided our first glimpse of the correlation. Rapoport (1975, translated 1982) was interested in the sizes of the geographical ranges of mammalian subspecies in tropical and extratropical latitudes and found that tropical subspecies had relatively smaller geographical ranges. Since Rapoport's pioneering work it has been shown that the mammalian pattern is mirrored by a wide variety of taxa (Stevens 1989). The ubiquity of the pattern suggests it is produced by an omni-

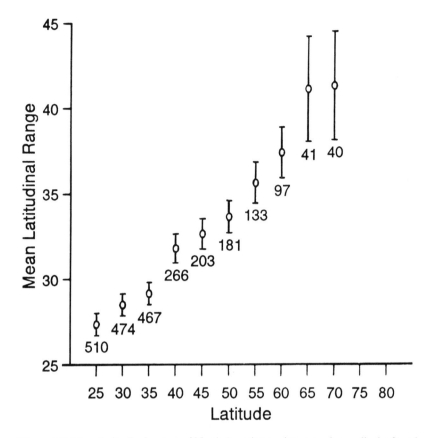

Figure 4.5 Mean latitudinal extent of North American existent marine mollusks found in different latitudes. Error bars are one standard error of the mean. Numbers below the bars are sample sizes.

present agent. This agent must operate on trees much in the same way it affects fish, reptiles, and marine mollusks. To understand how Rapoport's rule is produced in such a wide variety of organisms we need to step back and look at the big picture.

Global climatic patterns have the potential of influencing a wide range of organisms over a large spatial scale. The mild temperatures and sunny days of the tropics might lead one to suggest that the tropics are relatively benign and easy places to make a living. A little thought shows this to be an overly anthropocentric view. Even a quick stroll through a tropical forest ends any thoughts of how easy it is for plants to make a living or for animals to eat without being eaten. To understand the origin of this widespread misconcep-

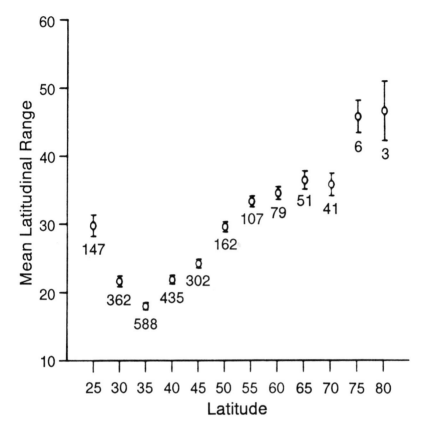

Figure 4.6 Mean latitudinal extent of North American freshwater and coastal fishes found in different latitudes. Error bars are one standard error of the mean. Numbers below the bars are sample sizes.

tion, ask a polar bear whether the tropics are benign. How an organism would define comfort and ease depends on the type of habitat from which it comes. We may view the tropics as comfortable because the tropics are the latitudinal zone of our origin. What seems comfortable to us would cause distress to many organisms adapted to other conditions.

The aspect of the climate that probably produces Rapoport's rule is not the climatic averages but rather the range of climatic conditions experienced during an individual's lifetime at different latitudes. On a summer's day in Fairbanks, Alaska, the temperature can reach into the 90s (30s on the Celsius scale), much like those temperatures found in a lowland Rain Forest in Costa Rica, but the winter temperatures of the two sites are very different. This

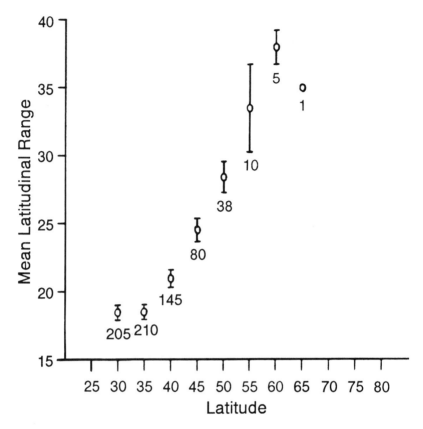

Figure 4.7 Mean latitudinal extent of North American reptiles and amphibians found in different latitudes. Error bars are one standard error of the mean. Numbers below the bars are sample sizes.

means that ravens, spruce trees, fish, and others that live at high latitudes must be able to tolerate or anticipate a relatively wide range of climatic conditions (temperature as well as other aspects of climate; see Stevens 1989 for details) to exploit any given habitat. Tropical species are not forced to cope with such a wide range of environmental conditions.

As a result, natural selection has produced relatively broad climatic tolerances (and therefore large geographical ranges) in high-latitude species and relatively narrower climatic tolerances (and so smaller ranges) in tropical ones. The ecological range of an individual of a high-latitude species must be very wide just to survive a single seasonal cycle. In contrast, individuals of tropical species with slightly larger ecological ranges than the norm would not be especially favored by natural selection in the tropics. To them there is no

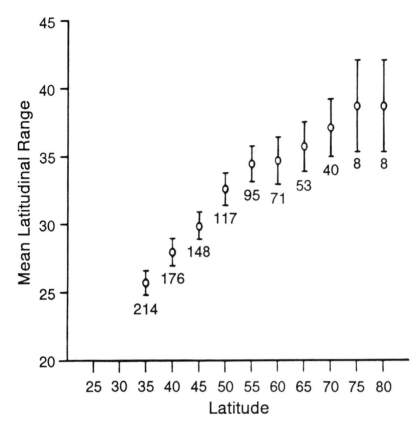

Figure 4.8 Mean latitudinal extent of North American mammals found in different latitudes. Error bars are one standard error of the mean. Numbers below the bars are sample sizes.

immediate advantage to a slightly broader range of tolerances when those tolerances allow for the exploitation of habitats only outside the range of experience.

This argument may sound vaguely reminiscent of the more familiar "narrower niches allow greater coexistence in the tropics" hypothesis, but one needs to be careful. No suggestion is being made that tropical species have narrower niches, exist in a system dominated by biotic rather than environmental constraints, or have evolved in a more stable environment. These points are semantic ones and the interested reader is referred to Stevens (1989) for a detailed discussion of them. To illustrate the difficulty, consider what effect Rapoport's rule is likely to have on the competitive environment of low-latitude communities.

The conceptual inconsistency hidden in the semantics can be grasped quickly through a brief digression into a review of resource utilization curves (figure 4.9). These curves are often presented in introductory ecology classes to illustrate how competition may evolve. On these plots the x axis orders available resources along some meaningful dimension. An often used example is the imaginary case of two birds species feeding on insects of different sizes. In that case the x axis might rank the insect prey according to size. The area under the two curves is the amount of resource necessary to maintain the species in the community and covers the range of resources the species uses. In figure 4.10, the curves have been redrawn but this time show greater specialization for one of the two species. The height of the curve changes because if the species uses a narrower range of resources it makes up for the loss by increasing its dependence on each of these resources.

The "narrower niches" hypothesis for greater species richness in the tropics proposes that more species are packed into tropical systems by narrowing the range of resources used by each (figures 4.11 and 4.12). Contrast this with what the existence of the Rapoport phenomenon suggests. If climate were

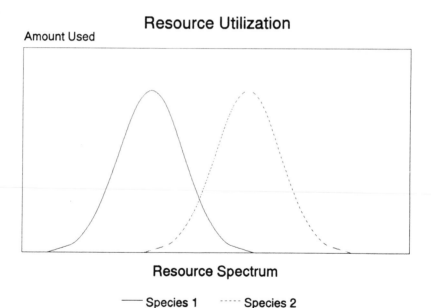

Figure 4.9 Hypothetical relationship between the resources used by two species. The ordinate scales the units of resources consumed by the given species at the sample site, the abscissa arranges the resources along some meaningful dimension (size, digestion difficulty, location within the habitat, etc.)

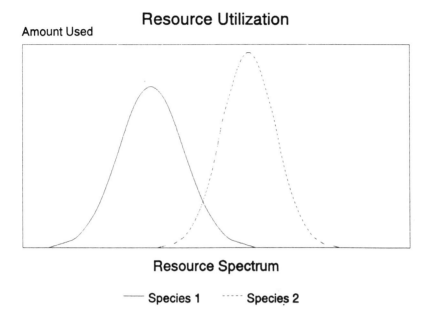

Resource Utilization

Amount Used

Resource Spectrum

——— Species 1   ····· Species 2

Figure 4.10 Another possible relationship between the resources used by two species. In contrast to figure 4.9, one species shown here is more specialized than the other.

somehow placed on the x axis, then tropical habitats would have a greater range of distinctive climate types because the seasonal cycle does not blur the boundaries of each as much. The change in the units of the x axis does not necessarily mean more species can be packed together, because each species still requires some minimum amount of resources to survive (noted before in reference to the taller curve of the specialist in figure 4.10). Tropical species would be forced to widen their niches over the more narrowly defined resource spectrum to make up for the smaller fraction of the total each distinctive resource comprises. Can one unambiguously say that tropical species are more specialized if the units of measure change?

The inconsistency in our thinking about the niche arises from the tendency of some ecologists to argue that the niche exists independent of the organism while others defend the view that the niche can be understood only as the resources an organism uses. If niches can be measured independently of the organisms, then the niches of tropical organisms are smaller, but the range of niches available to the community is shorter. Therefore tropical areas need not necessarily support more species because of their narrow niches. If instead we allow the organism to define the niche independent of the environment, then tropical niches are broader than extratropical ones because they cover more

Resource Utilization

Amount Used

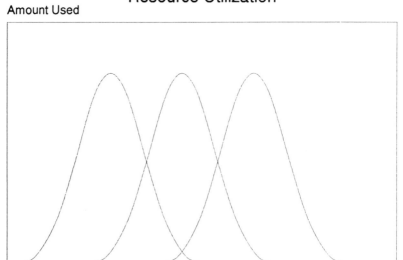

Resource Spectrum

Resource Utilization

Amount Used

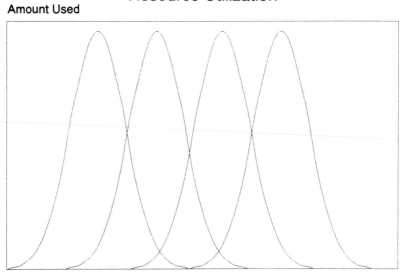

Resource Spectrum

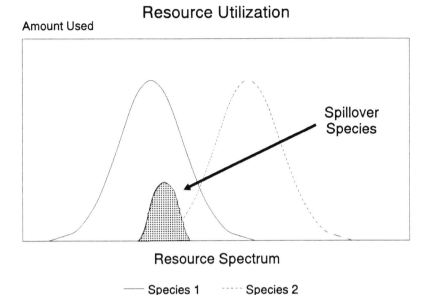

Resource Spectrum

―― Species 1 ┈┈ Species 2

Figure 4.13 Resource utilization curves for three species, the smallest of the three being a spillover species.

distinctively different climatic types. Once again there is no conceptual need for more species to be supported in tropical lands by virtue of the width of their niches. Some evidence has been accumulating that tropical species may be more generalized than their extratropical counterparts (reviewed by Price 1984).

The increase in species richness in low-latitude communities is produced by a completely different aspect of the Rapoport phenomenon, the living dead. Notice two things in the graph that describes this situation (figure 4.13). First notice that the area under the resource utilization curve of the living dead is less than that of the other species. This is because these spillover species are not successfully maintaining themselves just on the resources found within the space described by the resource utilization plot. The living dead, by definition, are in the site only because their parents were successful nearby. It is

FACING PAGE:
Top: Figure 4.11 The "packing" of a broad-niched species. Contrast with figure 4.12.
Bottom: Figure 4.12 The "packing" of narrow-niched species. Contrast with figure 4.11.

very easy to be taken by the simplicity of the resource utilization plots and not notice that they assume the community is a closed system. The dynamics of the living dead at treeline are the best antidote for this kind of thinking.

The second thing to notice is that the living dead can overlap heavily with resident species yet not be competitively excluded. Once again, this is because the living dead are decoupled from local conditions. A community with many living dead spills over the competitive limits we have grown to expect from the study of resource utilization curves and simple communities. If this view of nature is correct, one would expect that competition would be much harder to document in tropical communities because the living dead would dilute its effects. There is some support for this prediction.

Taking Connell's review (1983) of well-balanced, statistically sound studies of competition as a data base, one finds that over a wide range of taxa, competition is found relatively less often in tropical latitudes than in extratropical areas (figure 4.14). The opportunities for detection are more frequent, but competition is found less in low altitude communities. Even so, this should be considered very weak evidence because of the nonindependence of

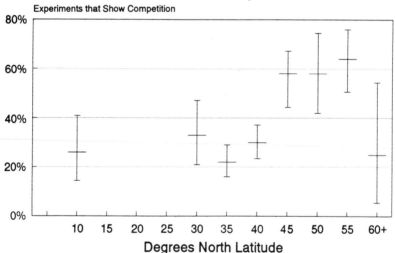

**Latitudinal Gradient
Incidence of Competition**

Experiments that Show Competition

Compiled by Connell 1983

Figure 4.14 The relative frequency of evidence of competition in pairwise species comparisons at different latitudes. Data are taken from Connell 1983. The statistical significance of the correlation is questionable owing to problems of nonindependence of the data points.

the studies and their multiple, nonindependent attempts to detect competition at each latitude. Nevertheless, the breadth of taxa included in the survey and the strong tendency for the detection of competition to be a function of latitude make this example worth considering.

A much more rigorous test of the prediction that communities at lower latitudes have more living dead than those found at higher latitudes comes from a latitudinal comparison of woodboring insect communities. Biogeographers have noted that oceanic islands of different sizes differ in the number of species they support. Small islands have smaller floras and faunas than larger islands. Biologists who study insects and plants have noticed an analogous pattern in their studies of insect feeding patterns. In any given site, abundant plants tend to have more associated species than rare plants have. This pattern is explained as the outcome of the limited resources rare plants provide as compared with the much greater wealth of resources contained in a larger population of plants. With greater resources, more herbivores can be supported.

The "host plants as islands" pattern can be seen when the number of woodboring insects to use a plant species in a site is plotted against the abundance of the host in that site (figure 4.15). The strength of this relationship is not a

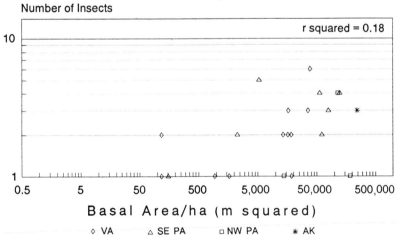

Figure 4.15 The relationship between the abundance of host tree species in four North American sites and the number of branch-inhabiting wood-boring insect species supported by each host. A more detailed analysis of these data can be found in Stevens 1986.

constant however. When a more southern, more species-rich site is added to those north temperate sites already plotted, the overall correlation begins to disappear (figure 4.16). This is because of several anomalous rare plants in the low-latitude site that support far more species than expected on the basis of the plant's abundance. The number of species using these rare plants exceeds the limits exhibited by other species. The weakening correlation between host abundance and herbivore species richness suggests spillover species are becoming more common at lower latitudes. The insects maintained in a site on very rare plants are likely to have moved into the site from neighboring areas where the host plant is more common.

The increasing irrelevance of our standard ways of thinking about communities with decreasing latitude is the major new insight derived from Rapoport's rule. Competitive interactions appear to be less important in limiting species richness at low latitudes. One of the central tenets of island biogeography theory, the relation between resources and the number of species using those resources, appears to find less and less support in continental communities of decreasing latitudes.

There are two sets of conclusions that can be drawn from these findings. For those concerned with the limits to species richness the message is to break out of the temporal dichotomy forced on us by the debate we inherited. Think of how space, rather than time, might better structure the discussion. Would

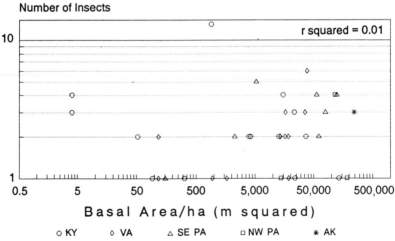

Figure 4.16 As in figure 4.15 but with an additional, more southern site.

it not be interesting to know how the shifting position of the continents has altered the latitudes at which different taxa are found and how those shifts have changed the species richness of the taxa? How does the geographical extent of higher taxonomic units correlate (if at all) with the number of species they contain? Do changes in the geographical range of taxa through time correlate with rates of speciation during that period or with the latitudes at which the taxa are found? It is through questions of this type that a better understanding of the direct role of latitude on speciation will be found.

The second set of conclusions focuses on the question of how best to preserve natural variety and what natural history museums can do. The existence of the living dead means that although patches of natural lands can be effectively managed in the high-latitude temperate and polar regions, the loss of species from tropical patches will not stop soon. There are fundamental differences in the organization of communities of different latitudes. Even if the pressures for deforestation and hunting were to be magically reduced, species extinctions would continue. The living dead would be the first to go and the most missed by the general public. The living dead are the exotics, the species with very limited geographical ranges and population sizes, the ones that most often become "poster children" for the environmental movement.

The challenge of maintaining the living dead is a political one. To limit the losses due to the disruption of spillover from source areas where a species does well, the size of preserves must be as large as politically possible. The political pressure for limiting conservation in the tropics is primarily local and with good reason. It is the local people who suffer most from the loss of immediate economic return on conserved lands. When trying to guarantee a long-term return on the land (through increased tourism) it is foolish to bet on the living dead as the main attraction. They may not be there when most needed.

Museums and other educational institutions need to be more adept at generating enthusiasm for actually visiting a tropical habitat. Gone are the days when the museum was the only way for the general public to see artifacts of exotic places. Now the tropics are only a couple of hundred dollars away by air. This new reality justifies a more mature approach to answering the question of why conserve tropical habitats. Whether or not people act on their enthusiasm is not as important as letting them in on why biologists find the tropics so interesting. Without diverse tropical forests we would have nothing to compare with the drab, impoverished forests we call our own.

REFERENCES

Allen, P. H. 1956. *The Rain Forests of Golfo Dulce.* Stanford: Stanford University Press.

Arno, S. R. 1984. *Timberline, Mountain, and Arctic Forest Frontiers*. Seattle: The Mountaineers.

Brockman, C. F. 1968. In H. S. Zim, ed. *Trees of North America: A Field Guide to the Major Native and Introduced Species North of Mexico*. New York: Golden Press.

Brown, J. H. 1988. Variation in desert rodent guilds: Patterns, processes, and scales. In J. H. R. Gee and P. S. Giller, eds., *Organization of Communities*, pp. 185–203. Oxford: Blackwell.

Brown, J. H. and A. C. Gibson. 1983. *Biogeography*. St. Louis: Mosby.

Cook, R. E. 1969. Variation in species density of North American birds. *Systematic Zoology* 18:63–84.

Connell, J. H. 1983. On the prevalence and relative importance of interspecific competition: Evidence from field experiments. *American Naturalist* 122:661–696.

Croat, T. B. 1978. *Flora of Barro Colorado Island*. Stanford: Stanford University Press.

Eldredge, N. and J. Cracraft. 1980. *Phylogenetic Patterns and the Evolutionary Process, Method, and Theory in Comparative Biology*. New York: Columbia University Press.

Holdridge, L. R. and L. J. Poveda A. 1975. *Arboles de Costa Rica*, vol. 1. *Palmas, Otras Monocoteledneas Arboreas, y Arboles con Hojas Compuestas o Lobuladas*. San José, Costa Rica: Centro Cientifico Tropical.

Pianka, E. R. 1966. Latitudinal gradients in species diversity: A review of concepts. *American Naturalist* 100:33–46.

Price, P. W. 1984. Alternative paradigms in community ecology. In P. W. Price, C. N. Slobodchikoff, and W. S. Gaud, eds., *A New Ecology: Novel Approaches to Interactive Systems;* pp. 353–383. New York: Wiley.

Pulliam, H. R. 1988. Sources, sinks, and population regulation. *American Naturalist* 132:652–661.

Rapoport, E. H. 1975. *Areografa: Estrategias Geograficas de las Especies*. Mexico City: Fondo de Cultura Econmica.

Rapoport, E. H. 1982. *Areography: Geographical Strategies of Species*, vol. 1. 1st English ed., B. Drausal, translator Publicacion Fundacin Bariloche. New York: Pergamon.

Simpson, G. G. 1964. Species density of North American recent mammals. *Systematic Zoology* 13:57–73.

Stevens, G. C. 1986. Dissection of the species-area relationship among wood-boring insects and their host plants. *American Naturalist* 128:35–46.

Stevens, G. C. 1989. The latitudinal gradient in geographical range: How so many species coexist in the tropics. *American Naturalist* 133:240–256.

Van Velzen, A. and W. T. Van Velzen, eds. 1988. The 88th Christmas Bird Count. *American Birds* 42:566–1163.

5 : Explaining Patterns of Biological Diversity: Integrating Causation at Different Spatial and Temporal Scales

Joel Cracraft

Patterns of diversity vary across space and time. Several billion years ago, there were very few species; today there are untold tens of millions. And over these billions of years, global patterns of species diversity have changed dramatically (Raup 1976; Bambach 1977; Sepkoski et al. 1981; Signor 1982; Niklas, Tiffney, and Knoll 1983; Padian and Clemens 1985; Valentine 1985; Benton 1989). Likewise, at any single point in time, such as the present, species diversity varies spatially, with latitudinal and longitudinal gradients observed across terrestrial and oceanic environments (Fischer 1960; MacArthur 1965, 1972; Whittaker 1972, 1977; Cracraft 1985).

How might these spatial and temporal patterns be explained? Many biologists have attempted to understand the causal dynamics of biological diversification, yet because of the complexity of the problem no general theory has emerged. Across large spatial and temporal scales, change in diversity is a first-order function of varying rates of speciation and extinction, and it therefore seems reasonable to infer that a general theory of diversification will be built around the causal dynamics of these two processes.

A Model of Speciation and Extinction Rate Controls

Many different factors have been implicated in the control of speciation and extinction rates. Changes in these rates define patterns of diversification, including (1) temporal changes in diversity within monophyletic groups, (2) temporal change in the diversity of the earth's biota, and (3) temporal and spatial changes in diversity across community assemblages, ecosystems, or biotas (Eldredge and Cracraft 1980; Cracraft 1985). Understanding the dynamics of these rate controls is, therefore, critical to developing a theory that explains how gradients of diversity are structured across space and time. Most neontologists and paleobiologists have directed their efforts toward explaining specific instances of these three types of pattern. As important as these case studies are, in order for a general theory of diversification to be developed we will need to identify those few, critically important rate controls of speciation and extinction that explain most of the observed variance within each of these different patterns. A single factor would not be expected to explain all the observed variance; neither would we expect each of the many factors that have been identified to have equal explanatory value. What can be hoped for, however, is that most of the observed variance in speciation rates from one clade to another, or among biotas, can be accounted for by a relatively small subset of factors and that variance in extinction rates will likewise be ascribed to a small subset of controls. It is from such an analysis that a general theory of diversification is likely to emerge.

Figure 5.1 presents a hypothesis for the rate control of speciation and extinction. It attempts to identify, and link together in a causal network, most of the processes that have been postulated to influence speciation and extinction rates. Each of these processes involves entities that are bounded in space and time, thereby making the processes themselves scale dependent (see also Wiens 1989). Thus, biotic interactions or demographic changes associated with population growth rates operate on spatial and temporal scales much smaller than, say, geomorphological or climatic changes. These scale effects become a consideration in evaluating the relative contribution of these processes to variation in speciation and extinction rates. Thus, organisms of disparate body sizes and life history parameters (e.g., generation time) comprise populations that generally exhibit different demographic characteristics and sizes of distributions (Brown and Maurer 1987; Gaston 1990). Such differences in spatial scale may, for example, have profound effects on rates of speciation or extinction. If so, then generalities that are formulated to explain rates of speciation or extinction in large-bodied vertebrates might not apply to small-bodied arthropods whose distributions are often a fraction the size of those of vertebrates. It is difficult to know how these differences in scale affect the generality of hypotheses about rate controls, and this ignorance is likely

A DIVERSITY-INDEPENDENT MODEL OF BIOTIC DIVERSIFICATION

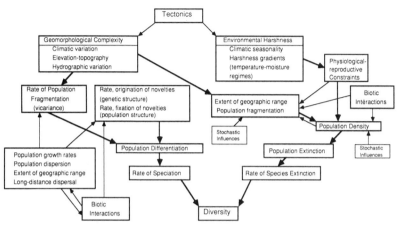

Figure 5.1 A diversity-independent model of biotic diversification. This flow diagram identifies the major rate controls of speciation and extinction. Diversification is postulated to be diversity independent because the most important rate controls (heavy arrows) are not influenced by standing diversity. See text.

to continue until much more information is obtained on the systematics and historical biogeography of groups of small-bodied taxa. Some of the controversies about the causal mechanics of diversification have arisen because investigators have emphasized processes at one scale while deemphasizing or ignoring processes at different scales. The view of diversification outlined in figure 5.1 attempts to integrate processes at different scale because only through such an effort will we eventually be able to critically evaluate the role each plays in determining speciation and extinction rates.

The conceptualization of figure 5.1 is descriptive of a *diversity-independent model of diversification;* that is, it is postulated that standing diversity does not have a significant causal influence on those factors controlling speciation and extinction rates. This is not to say *diversity-dependent* effects might not exist, only that they are suggested to have a negligible influence on these rates. Further commentary on the relative importance of diversity-dependent effects will be presented in a subsequent section.

The Rate Controls of Speciation

There are just three primary determinants of the rate of speciation (figure 5.1). Two of these might be termed *intrinsic* in the sense that they are characteristic of individual organisms and populations. These intrinsic processes are necessary and sufficient for speciation to occur. The first includes processes affecting the rate at which evolutionary novelties are introduced into the ontogenies of individual organisms. These novelties include mutational events (of any kind) within the genome and their ontogenetic/phenotypic effects on the organism. The second intrinsic factor is the set of processes influencing the rate at which those novelties become fixed in populations. All things being equal, the more often variants enter a population and the more rapidly they become fixed, the higher the rate of taxonomic differentiation.

These two sets of processes define the necessary mechanistic context of taxonomic differentiation, or speciation. Species, regardless of how they might be defined, represent populations that have differentiated. One or more evolutionary novelties have arisen and spread through the population to become fixed, thus allowing that population, or lineage, to be diagnosed as being distinct from its close relatives.

The third primary determinant of speciation rate is *extrinsic* in that it largely involves processes external to the population that is differentiating. Spatial and temporal variation in large-scale environmental heterogeneity, or geomorphological complexity, influences the rate at which populations become isolated and therefore differentiate. In the sense used here, geomorphological complexity refers to the array of environmental characteristics such as variation in climate, topography, and river systems that are the result of geological evolution. That there should be a predicted causal connection between changes in earth history and the rate control of speciation follows from two propositions: (1) speciation is allopatric in most groups of organisms and relies upon the origin and maintenance of geographical and/or ecological barriers (Mayr 1963; Grant 1971; Bush 1975; White 1978); and (2) the rate of formation of these barriers is a function of the rate change in large-scale geomorphological evolution, which is a manifestation of the inherent increase in complexity of the lithosphere coupled with the degradation of that complexity by erosional processes.

Geographic isolation is widely regarded as a primary factor in promoting taxonomic differentiation within the majority of marine and terrestrial organisms. Evidence that nonallopatric differentiation is of frequent occurrence in animals is limited (Mayr 1963; Futuyma and Mayer 1980), and even in plants, where polyploidy is common in many groups, allopatric disjunction seems to be a frequent mode of speciation (Grant 1971; the relative frequency of different modes of speciation within many plant, and animal, groups is still an open question, however; see Lynch 1989).

Allopatric speciation can either involve dispersal of propagules across a preexisting barrier (peripatric differentiation), followed by differentiation, or it can result from vicariance of a widespread ancestral species by a newly arisen barrier (dichopatric differentiation). In either case, isolation and differentiation rely upon the origination and maintenance of geomorphological or ecological barriers.

Irreversible flows of energy and matter into the lithosphere drive mantle convection and thus create large-scale patterns of geomorphological change. Like all historical systems, the lithosphere possesses an inherent tendency to increase in complexity (at least as long as matter-energy flows are maintained), and our perception of that tendency is reflected in our ability to reconstruct that history. Geological processes degrading complexity, and consequently obscuring historical patterns, generally function over smaller spatial scales than those processes generating that complexity. Thus, plate tectonic events can lead to environments of high structural complexity (e.g., along active plate margins) and also have a major influence on climatic change. Both, in turn, directly underlie the spatial and temporal history of geographic and ecological isolating barriers (e.g., Vrba 1985). Secular variation in the rate of barrier formation is independent of the diversity present in any given area, and other things being equal, the rate of speciation should be related to the rate of barrier formation.

The causal nexus of speciation rate controls is more complex than the three sets of processes just considered (see figure 5.1). The rate of population fragmentation in response to changes in geomorphological complexity is also influenced by processes operating within and among populations. The degree to which particular populations become fragmented by any given barrier may also be a function of how individuals of that population respond ecologically and behaviorally to that barrier, the demographic structure of the population, and the context of its interactions with populations of other species. Thus, the manner in which populations distribute themselves in space (e.g., factors of population dispersion, long-distance dispersal, and extent of geographic range) and the factors that regulate population densities across their range will play a role in determining the magnitude of the effects of changes in geomorphological complexity. Likewise, any process that influences the demographic structure of populations will have a direct effect on the rate at which novelties will become fixed in those populations.

These considerations suggest that speciation rate is a function of numerous processes each having its own spatial and temporal context. As will be discussed below, differences in spatial and temporal scale can provide insight into the relative importance of these processes in having major influences on speciation rate.

The Rate Controls of Extinction

Understanding the causal nexus of extinction is much more difficult than that of speciation because the former involves a multitude of interacting processes whose relative importance is generally not easy to establish. And although a hierarchy of causal effects, from "ultimate" to "proximate," can be postulated, extinction is at base very simple: it is the statistical summation of all the individual organisms in all the populations of a species dying without replacement. Yet, despite the fact that extinction has in some sense its causal locus at the level of the individual organism, events leading to extinction are initiated at many spatial and temporal scales. The main question addressed here is which kinds of factors are likely to be important in establishing large-scale patterns of diversity.

Theories of extinction and causal analyses of case studies are numerous, and no attempt will be made to review the extensive literature (see, for example, Nitecki 1981, 1984; Berggren and Van Couvering 1984; Holland and Trendall 1984; Martin and Klein 1984; Elliott 1986; Stanley 1987; Larwood 1988; Donovan 1989). The model shown in figure 5.1 describes the major causal pathways generally implicated in extinction. Our focus here is to identify "ultimate" causation of large-scale patterns of extinction. Since extinction involves the death of individual organisms, there is often a tendency to emphasize "proximate" causes such as local biotic interactions and/or stochastic effects. Even though this might speak to the importance of scale in assembling a complete calculus of extinction, it begs the issue of accounting for large-scale patterns where the identification of "ultimate" causation is biologically more interesting.

Two important pathways of "ultimate" causation are postulated in the model of figure 5.1. Both are causal functions of earth history, coupled with the earth's rotational dynamics. As noted earlier, plate tectonic events create geomorphological complexity, which forms the basis for changes in the rate of population fragmentation. Although this large-scale environmental heterogeneity facilitates allopatric speciation, reduced geographic ranges and fragmented populations result in reduced population sizes and therefore higher probabilities of extinction (Jackson 1974; Fowler and MacMahon 1982; Brown and Maurer 1987; Jablonski 1987; but the relationship between fragmentation and extinction is complicated: see Quinn and Hastings 1987). Large-scale tectonic activity has also been implicated in the rate control of extinction through the direct elimination of habitat, including, for example, the loss of shallow marine environments following continent-continent collision (Valentine and Moores 1970, 1972; Valentine 1971, 1973), changes in sealevel (Newell 1967; Schopf 1974; Hallam 1989), and volcanism (Axelrod 1981).

Tectonic processes affect extinction rates through a second causal pathway, namely, large-scale climatic change. The earth's climate over time is primarily, but not exclusively, a function of the earth's rotational dynamics, its tectonic milieu (relative position of the oceans and landmasses, extent and location of topographic relief, and so on), and the changing nature of the biosphere itself (e.g., its biomass and where it is distributed). Change toward extreme conditions of environmental harshness, especially extremes of temperature and moisture, is said to be a major factor leading to an increase in extinction (Axelrod 1967; Valentine 1973; Galloway and Kemp 1981; Kemp 1981; Hickey 1981, 1984; Kauffmann 1984; Stanley 1984, 1987, 1988; to name only a few). Why this should be so can be inferred from several well-supported observations. First, change in population density (birth rates minus death rates) for all organisms is influenced by alterations in environmental favorableness (here taken to be measured by temperature and moisture regimes). Change toward harsher, less favorable environments results in lower birth rates and/or increased death rates (e.g., Richerson and Lum 1980; Osmond et al. 1987). Second, most kinds of organisms are narrowly distributed and are found along temperature and moisture gradients that, in absolute terms, can be characterized as being "favorable" relative to the extremes of these gradients. This is a consequence of most organisms' sharing similar physiological-biochemical properties, on the one hand, and of the difficulty most organisms have, physiologically and biochemically, in adjusting to absolute harsh environments, on the other. Evolutionarily, most kinds of organisms have failed to solve the effects of these environmental extremes. This means that as climate changes over time toward harsher environments, organisms will generally shift their distributions toward more favorable environments (Huntley and Webb 1989), or if that is not possible, their populations will go extinct.

Two other sets of more "proximate" causes of extinction have been widely discussed. Biotic interactions—primarily competition and predation—are often proposed as important causes of extinction at both large (e.g., Van Valen and Sloan 1977; Huston 1979; Kauffman 1984; Knoll 1984) and small (Bengtsson 1989; and many others) spatiotemporal scales. The general effects of these processes on population densities are straightforward; their effects on species extinction are more problematical and will be discussed further below. A second set of "proximate" causes includes stochastic processes that result from random, uncertain events. Stochastic causes of extinction are much more likely to be important in species extinction when population densities become low. Shaffer (1987:71) categorizes these events into four main groups, including demographic uncertainty associated with survival and reproduction, environmental uncertainty associated with weather and biotic interactions, natural catastrophes such as fires and floods, and genetic uncertainty resulting from genetic drift or inbreeding (see also Shaffer 1981; other papers in Soulé 1987).

Diversity-Dependent Diversification: The Effects of Processes Must Scale Upward

Most current conceptualizations of diversification are described in terms of *diversity-dependent equilibrial dynamics*. Speciation and extinction rates are postulated to be regulated by factors whose influences are dependent upon the standing diversity of the system relative to some equilibrial value (Webb 1969, 1976; Rosenzweig 1975; Valentine 1977; Sepkoski 1978, 1979, 1984; Carr and Kitchell 1980; Rosenzweig and Taylor 1980; Marshall et al. 1982). If standing diversity exceeds the equilibrial value, or carrying capacity of the system, the speciation rate declines and/or the extinction rate increases. In contrast, when diversity within a system is below the carrying capacity, the speciation rate increases and/or the extinction rate declines.

Diversity-dependent models are explicit conceptual analogues of density-dependent, population regulation models, especially as they have been applied to island biogeographic theory (e.g., MacArthur and Wilson 1963, 1967; MacArthur 1969) but that are extended to the higher hierarchical levels of species and clades (Valentine 1972:213; Rosenzweig 1975; Sepkoski 1978, 1979, 1984; Hoffmann and Kitchell 1984: 26; Maurer 1989). In effect, the biosphere is modeled as a large population bottle: resources are finite and biological interactions within the system determine the quantitative expression of birth (speciation) and death (extinction) processes. Often one of the more fundamental premises of diversity-dependent models is that the system under consideration has an upper limit to its biomass, which in turn sets a limit to the numbers of individual organisms present in the system and, consequently, the maximal number of allotted species (MacArthur 1965, 1969; Margalef 1972; Simberloff 1974; Levinton 1979). Hence, an implicit characteristic of the logical structure of these models is that processes regulating speciation and/or extinction are expressions of (or reducible to) ecological processes acting among individual organisms in local communities.

There is considerable debate within contemporary ecology about the importance of biotic interactions such as competition and predation in structuring the distribution and abundances of species populations and therefore communities. These are processes that have spatial and temporal scales at the level of local populations, and their effects, on birth and death rates, are also at the level of local communities. A necessary requirement of diversity-dependent diversification theory is not only that the effects of these population-level processes scale upward to the level at which speciation and extinction are important but that the intensity of these processes be inversely related to standing diversity. If competition is to have an effect on extinction, for example, competitive interactions must extend across the distributions of the competing species and have sufficient longevity to result in extirpation of all populations of one of the species. This means, also, that the distributions of the competing species must be largely or entirely sympatric, the competition coefficients must be of similar magnitude and direction spatially and temporally, and the variation in other biotic and abiotic factors across the distributions of these species must have no material influence on the competitive outcome.

Arguments for diversity-dependent diversification have failed to consider this scale effect, yet it is critical to this model of speciation and extinction rate control. Even though biotic interactions may be important, even crucial, in structuring local communities, one cannot therefore conclude that the effects of those processes—that is, changes in the relative abundances of populations—will have any necessary consequences at the spatial and temporal scales of species (see Wiens 1989 for a general discussion about scale effects of ecological processes). In fact, there is little empirical evidence that they do (Benton 1987).

Diversity-Independent Diversification: A Comparison of Rate Controls

Geomorphological (including topographical) complexity has long been appreciated as an important factor in promoting speciation by enhancing the likelihood of isolation via vicariance or long-distance dispersal (e.g., Simpson 1964; Ross 1972), but the generality and implications of the correlation between complexity and diversity are in need of much more work (see Cracraft 1985; Vrba 1985 for further discussion). A causal relationship between changes in earth history and speciation can be examined directly by the methods of vicariance biogeography (Platnick and Nelson 1978; Nelson and Platnick 1980; Humphries and Parenti 1986), which also provide a means of relating variation in geomorphological complexity to speciation rate (figure 5.2). Evidence is accumulating that species-level components of biotas show historical congruence in their biogeographic patterns (Rosen 1978, 1979; Cracraft 1986; Cracraft and Prum 1988), as would be predicted if changes in earth history established the pattern and tempo of geographic isolation (see also Lynch 1989).

Whereas the rate of population fragmentation is obviously a potentially important factor in explaining variation in patterns of diversity among clades as well as within entire biotas (figure 5.2), the relative importance of factors influencing the rates of origination and fixation of novelties is less well understood. The extent of genetic variation within species has been implicated in their rate of speciation, although the evidence is ambiguous or contradictory (Avise 1977; Patton and Sherwood 1983). It may be very difficult, from a comparative standpoint, to separate the effects on speciation rate of increased

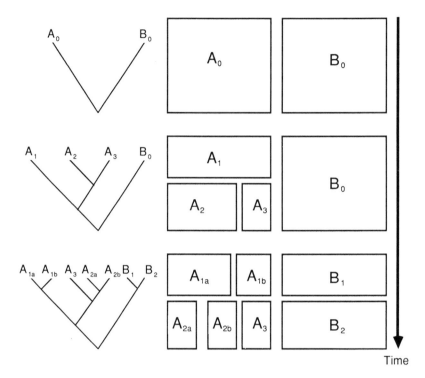

Figure 5.2 A model and test for the control of speciation rate by earth history. Two sister species A_0 and B_0 are components of two biotas that exhibit different rates of geomorphological evolution over time. If rates of vicariance are important in regulating speciation rate, and if it is assumed that there is no extinction or differentiation following long-distance dispersal, then the clade distributed in the most geomorphologically complex area will be characterized by more species. The model can be used to explain differences not only in diversity among lineages of a clade but also within and among biotas.

genetic variability, on the one hand, and of those factors promoting the rapid fixation of novelties, on the other. Thus, high karyotypic variability within populations might not translate into higher speciation rates unless coupled with demographic characteristics (e.g., population subdivision into isolated demes) that promote fixation (Wilson et al. 1975; Bush et al. 1977). The rate at which novelties are fixed may be very important in some groups for setting the tempo of speciation. Factors such as sexual selection, for example, have been implicated by a number of workers as an important mechanism increasing speciation rate (Lande 1981; West-Eberhard 1983; Carson 1986), but much more theoretical and empirical work will be necessary to establish this.

For reasons already given, it is unlikely that biotic interactions, in particular predation and competition, will function as effective causes of large-scale extinction patterns. Thus, whereas it may be possible to identify an individual example of an extinction event as being caused by predation or competition, the specific demographic and biological characteristics leading to that event would not be expected to apply across other species in a biota. And unless these characteristics can scale upward to the level of biotas, biotic interactions should play little, if any, role in explaining large-scale extinction patterns, either in space or time. Changes in earth history seem, therefore, to play the major role in instigating large-scale patterns of extinction. The effect of these changes can, however, be manifested over very different time scales (e.g., Crowley and North 1988). Some of these changes can act very rapidly to destroy or alter living space; the effects of other changes, on the other hand, may extend over long time periods, and consequently, measured extinction rates will be low. Large-scale changes toward harsher environmental regimes can have "direct" effects on organisms in that they challenge their physiological-biochemical ability to respond to these changes. Or those effects can be "indirect" in that they affect other organisms on which they are dependent. Vermeij (1989) argues, for example, that reductions in primary productivity, which are probably brought about by abiotic environmental change, have played a significant role in the extinction of consumer species (see also Fowler and MacMahon 1982).

The Pattern of Biospheric Evolution

Diversity-dependent diversification is said to operate in ecosystems or biotas at or near their carrying capacity, which implies an upper limit to biomass and, in some cases, even a (geological) long-term constancy of biomass. Although most proponents of diversity dependence accept, implicitly or explicitly, some notion of carrying capacity for global biomass, specific arguments in support of this assumption are rarely presented (a notable exception is Van Valen 1976). Benton (1979) reviewed the geochemical evidence for and against the hypothesis of constant biomass and concluded that positive evidence for constancy is lacking. Some data, instead, suggest a temporal increase in organic carbon (Jackson 1973, 1975), which may indicate an increase in productivity (Benton 1979). Quantitative comparisons are extremely difficult and subject to numerous biases, however, and virtually all conclusions are open to conflicting interpretations.

Three additional lines of argumentation favor the hypothesis that biomass has not been constant over geological time. The first is theoretical and describes an expectation. The components of the biosphere are thermodynamically structured far from equilibrium by the inflow of matter and energy (Johnson 1981; Mercer 1981; Brooks and Wiley 1988). Transduction of energy allows living organisms to break down inorganic matter and incorporate those products as increased biomass (although some organic matter is certainly recycled back to the lithosphere). Thus, on theoretical grounds (Mercer 1981; Brooks and Wiley 1988), the biosphere would be expected to exhibit an inherent tendency to increase in biomass unless constrained by physical factors originating "external" to the biosphere itself.

A second argument supporting an increase in biomass through time is empirical, and it is derived from observations about the Phanerozoic diversity curve. Diversity has been expanding exponentially for the last 250–300 m.y. That increase is seen for marine invertebrates (Raup 1976; Sepkoski et al. 1981; Signor 1982), vertebrates (Sepkoski 1981; Padian and Clemens 1985; Benton 1989), and plants (Knoll, Niklas, and Tiffney 1979; Niklas, Tiffney, and Knoll 1983; Knoll 1984). Omitted from these compilations, moreover, are terrestrial arthropods, especially insects, spiders, and mites, in addition to parasites of all kinds; almost certainly these taxa have exhibited a correlated pattern of increased diversity, particularly with that of plants.

This increase in diversity, it can be suggested, documents an increase in biomass, particularly at lower trophic levels, but also among consumers as well. Within a diversity-dependent viewpoint, the only alternative explanation would be that this increased diversity has been achieved at the expense of average relative abundance. No evidence seems to support this latter hypothesis.

A third line of evidence not only suggests that the Cenozoic high diversity has the potential to increase even further but at the same time identifies the primary external constraint on biomass evolution. Consider the following observations. About 99% of the world's biomass and 65–70% of its net primary productivity are found in terrestrial ecosystems (e.g., Krebs 1978:523). Of those figures, the greatest proportion (roughly 40% and 30%, respectively) is produced in tropical moist forests. Furthermore, these forests occupy only about 10% of the land surface area, while at the same time housing the most diverse terrestrial ecosystems.

The major constraint on global primary productivity at the present time is the steep latitudinal gradient in environmental harshness, particularly as manifested by a decrease in mean annual temperature toward higher latitudes (see discussion in Cracraft 1985). Abundant paleobotanical evidence suggests that when temperature (or moisture) extremes are ameliorated, vegetational associations capable of greater productivity expand geographically and replace the less productive associations of higher latitudes. This implies that world biomass would increase substantially if warm, equable climates were to become more widespread. The question whether there exists a theoretical upper limit to biomass is perhaps moot, given that evidence does not indicate such a limit has ever been attained. The potential for increased expansion of highly productive ecosystems into high-latitude temperate zones suggests that biomass is not in equilibrium but that it is constrained by extrinsic factors, primarily environmental harshness.

Within much of paleobiology, an equilibrial framework has placed conceptual constraints on the description of diversification: the biosphere consists of a finite resource base, and the problem is how that base is to be partitioned. How many species can be packed into the space? What will be their limiting similarities? How many niches are there and how many remain to be filled? Not addressed from this perspective is the evolutionary history of the resources themselves.

The arguments presented here suggest diversification should not be considered a problem of resource partitioning. Speciation expands the ecological dimensionality of communities and provides the basis for increased energy flow as species are added to the ecosystem. And because eventually speciation carries with it a change in form, and sometimes function, through time new morphologies come to exploit (i.e., dissipate) more and more energy and matter in novel ways, with the result that biomass can expand rather than remain constant (Flessa and Levinton 1975; Johnson 1981; Mercer 1981; Kitchell and Carr 1985). All biomass, moreover, had its origins in the matter of the lithosphere, and as long as the amount processed from the lithosphere is greater than the amount lost to it, the mass of the biosphere will continue to expand. The canonical pattern inherent in biospheric evolution is one of increasing complexity.

REFERENCES

Avise, J. C. 1977. Genic heterozygosity and the rate of speciation. *Paleobiology* 3:422–432.
Axelrod, D. I. 1967. Quarternary extinctions of large mammals. *University of California Publications in Geological Science* 74:1–42.
Axelrod, D. I. 1981. Role of volcanism in climate and evolution. *Geological Society of America Special Paper* no. 185:1–59.
Bambach, R. K. 1977. Species richness in marine benthic habitats through the Phanerozoic. *Paleobiology* 3:152–167.
Bengtsson, J. 1989. Interspecific competition increases local extinction rate in a metapopulation system. *Nature* 340:713–715.
Benton, M. J. 1979. Increase in total global biomass over time. *Evolutionary Theory* 4:123–128.
Benton, M. J. 1987. Progress and competition in macroevolution. *Biological Reviews* 62:305–338.

Benton, M. J. 1989. Patterns of evolution and extinction in vertebrates. In K. Allen and D. Briggs, eds., *Evolution and the Fossil Record*, pp. 218–241. London: Belhaven Press.

Berggren, W. A., and J. A. Van Couvering, eds. 1984. *Catastrophes and Earth History*. Princeton: Princeton University Press.

Brooks, D. R. and E. O. Wiley. 1988. *Evolution as Entropy*, 2d ed. Chicago: University of Chicago Press.

Brown, J. and B. A. Maurer. 1987. Evolution of species assemblages: Effects of energetic constrains and species dynamics in the diversification of the North American avifauna. *American Naturalist* 130:1–17.

Bush, G. L. 1975. Modes of animal speciation. *Annual Review of Ecology and Systematics* 6:339–364.

Bush, G. L., S. M. Case, A. C. Wilson, and J. L. Patton. 1977. Rapid speciation and chromosomal evolution in mammals. *Proceedings of the National Academy of Sciences* 74:3942–3946.

Carr, T. R. and J. A. Kitchell. 1980. Dynamics of taxonomic diversity. *Paleobiology* 6:427–443.

Carson, H. L. 1986. Sexual selection and speciation. In S. Karlin and E. Nevo, eds., *Evolutionary Processes and Theory*, pp. 391–409. New York: Academic Press.

Cracraft, J. 1985. Biological diversification and it causes. *Anuals of the Missouri Botanical Garden* 72:794–822.

Cracraft, J. 1986. Origin and evolution of continental biotas: Speciation and historical congruence within the Australian avifauna. *Evolution* 40:977–996.

Cracraft, J., and R. O. Prum. 1988. Patterns and processes of diversification: Speciation and historical congruence in some Neotropical birds. *Evolution* 42:603–620.

Crowley, T. J., and G. R. North. 1988. Abrupt climatic change and extinction events in earth history. *Science* 240:996–1002.

Donovan, S. K., ed. 1989. *Mass Extinctions: Processes and Evidence*. New York: Columbia University Press.

Eldredge, N. and J. Cracraft. 1980. *Phylogenetic Patterns and the Evolutionary Process*. New York: Columbia University Press.

Elliott, D. K., ed. 1986. *Dynamics of Extinction*. New York: Wiley.

Fischer, A. 1960. Latitudinal variations in organic diversity. *Evolution* 14:64–81.

Fowler, C. W. and J. A. MacMahon. 1982. Selective extinction and speciation: Their influence on the structure and functioning of communities and ecosystems. *American Naturalist* 119:480–498.

Flessa, K. W. and J. S. Levinton. 1975. Phanerozoic diversity patterns: Tests for randomness. *Journal of Geology* 83:239–248.

Futuyma, D. J. and G. C. Mayer. 1980. Non-allopatric speciation in animals. *Systematic Zoology* 29:254–271.

Galloway, R. W. and E. M. Kemp. 1981. Late Cainozoic environments in Australia. A. Keast, ed., *Ecological Biogeography in Australia*, pp. 51–80. Amsterdam: Dr. W. Junk.

Gaston, K. J. 1990. Patterns in the geographical ranges of species. *Biological Reviews* 65:105–129.

Grant, V. 1971. *Plant Speciation.* New York: Columbia University Press.

Hallam, A. 1989. The case for sea-level change as a dominant causal factor in mass extinction of marine invertebrates. *Philosophical Transactions of the Royal Society of London* 325B:437–455.

Hickey, L. J. 1981. Land plant evidence compatible with gradual, not catastrophic, change at the end of the Cretaceous. *Nature* 292:529–531.

Hickey, L. J. 1984. Changes in the angiosperm flora across the Cretaceous-Tertiary boundary. In W. A. Berggren and J. A. Van Couvering, eds., *Catastrophes and Earth History,* pp. 279–313. Princeton: Princeton University Press.

Hoffman, A. and J. A. Kitchell, 1984. Evolution in a pelagic plankic system: A paleobiologic test of models of multispecies evolution. *Paleobiology* 10:9–33.

Holland, H. D. and A. F. Trendall, eds. 1984. *Patterns of Change in Earth Evolution.* Berlin: Springer-Verlag.

Humphries, C. J. and L. R. Parenti. 1986. *Cladistic Biogeography.* Oxford: Oxford University Press.

Huntley, B. and T. Webb, III. 1989. Migration: Species' response to climatic variations caused by changes in the earth's orbit. *Journal of Biogeography* 16:5–19.

Huston, M. 1979. A general hypothesis of species diversity. *American Naturalist* 113:81–101.

Jablonski, D. 1987. Heritability at the species level: Analysis of geographic ranges of Cretaceous mollusks. *Science* 238:360–363.

Jackson, J. B. C. 1974. Biogeographic consequences of eurytopy and stenotopy among marine bivalves and their evolutionary significance. *American Naturalist* 108:541–560.

Jackson, T. A. 1973. "Humic" matter in the bitumin of ancient sediments: Variations through geologic time. *Geology* 1:163–166.

Jackson, T. A. 1975. "Humic" matter in the bitumin of pre-Phanerozoic and Phanerozoic sediments and its paleobiological significance. *American Journal of Science* 275:906–953.

Johnson, L. 1981. The thermodynamic origin of ecosystems. *Canadian Journal of Fisheries and Aquatic Science* 38:571–590.

Kauffman, E. G. 1984. The fabric of Cretaceous marine extinction. In W. A. Berggren and J. A. Van Couvering, eds., *Catastrophes and Earth History,* pp. 151–246. New York: Princeton University Press.

Kemp, E. M. 1981. Tertiary palaeogeography and the evolution of Australian climate. In A. Keast, ed., *Ecological Biogeography in Australia,* pp. 31–49. Amsterdam: Dr. W. Junk.

Kitchell, J. A. and T. R. Carr. 1985. Nonequilibrium model of diversification: Faunal turnover dynamics. In J. W. Valentine, ed., *Phanerozoic Diversity Patterns: Profiles in Macroevolution,* pp. 277–309. Princeton: Princeton University Press.

Knoll, A. H. 1984. Patterns of extinctions in the fossil record of vascular plants. In M. Nitecki, ed., *Extinctions,* pp. 21–68. Chicago: University of Chicago Press.

Knoll, A. H., K. J. Niklas, and B. H. Tiffney. 1979. Phanerozoic land-plant diversity in North America. *Science* 206:1400–1402.

Krebs, C. J. 1978. *Ecology: The Experimental Analysis of Distribution and Abundance.* New York: Harper and Row.

Lande, R. 1981. Models of speciation by sexual selection on polygenic traits. *Proceedings of the National Academy of Sciences* 78:3721–3725.

Larwood, G. P., ed. 1988. *Extinction and Survival in the Fossil Record.* Oxford: Oxford University Press.

Levinton, J. S. 1979. A theory of diversity equilibrium and morphological evolution. *Science* 204:335–336.

Lynch, J. D. 1989. The gauge of speciation: On the frequencies of modes of speciation. In D. Otte and J. A. Endler, eds. *Speciation and Its Consequences,* pp. 527–553. Sunderland, Mass.: Sinauer Associates.

MacArthur, R. H. 1965. Patterns of species diversity. *Biological Reviews* 40:510–533.

MacArthur, R. H. 1969. Patterns of communities in the tropics. *Biological Journal of the Linnaean Society* 1:19–30.

MacArthur, R. H. 1972. *Geographical Ecology.* New York: Harper and Row.

MacArthur, R. H. and E. O. Wilson. 1963. An equilibrium theory of insular zoogeography. *Evolution* 17:373–387.

MacArthur, R. H. and E. O. Wilson. 1967. *The Theory of Island Biogeography.* Princeton, N.J.: Princeton University Press.

Margalef, R. 1972. Homage to Evelyn Hutchinson, or why there is an upper limit to diversity. *Transactions of the Connecticut Academy of Arts and Sciences* 44: 213–235.

Marshall, L. G., S. D. Webb, J. J. Sepkoski, Jr., and D. M. Raup. 1982. Mammalian evolution and the Great American Interchange. *Science* 215:1351–1357.

Martin, P. S. and R. G. Klein, eds. 1984. *Quaternary Extinctions* (Tucson: University of Arizona Press).

Maurer, B. A. 1989. Diversity-dependent species dynamics: Incorporating the effects of population-level processes on species dynamics. *Paleobiology* 15:133–146.

Mayr, E. 1963. *Animal Species and Evolution.* Cambridge, Mass.: Harvard University Press.

Mercer, E. H. 1981. *The Foundations of Biological Theory.* New York: Wiley.

Nelson, G. J. and N. I. Platnick. 1980. *Systematics and Biogeography: Cladistics and Vicariance.* New York: Columbia University Press.

Newell, N. D. 1967. Revolutions in the history of life. *Special Papers of the Geological Society of America* 89:63–91.

Niklas, K. J., B. H. Tiffney, and A. H. Knoll. 1983. Patterns in vascular land plant diversification. *Nature* 303:614–616.

Nitecki, M. H., ed. 1981. *Biotic Crises in Ecological and Evolutionary Time.* New York: Academic Press.

Nitecki, M. H., ed. 1984. *Extinctions.* Chicago: University of Chicago Press.

Osmond, C. B., M. P. Austin, J. A. Berry, W. D. Billings, J. S. Boyer, J. W. H. Dacey, P. S. Nobel, S. D. Smith, and W. E. Winner. 1987. Stress physiology and the distribution of plants. *Bioscience* 37:38–48.

Padian, K. and W. A. Clemens. 1985. Terrestrial vertebrate diversity: Episodes and insights. In J. W. Valentine, ed., *Phanerozoic Diversity Patterns: Profiles in Macroevolution,* pp. 41–96. Princeton: Princeton University Press.

Patton, J. L. and S. W. Sherwood. 1983. Chromosome evolution and speciation in rodents. *Annual Review of Ecology and Systematics* 14:139–158.

Platnick, N. I. and G. J. Nelson. 1978. A method of analysis for historical biogeography. *Systematic Zoology* 27:1–16.

Quinn, J. F. and A. Hastings. 1987. Extinction in subdivided habitats. *Conservation Biology* 1:198–208.

Raup, D. M. 1976. Species diversity in the Phanerozoic: A tabulation. *Paleobiology* 2:279–288.

Richerson, P. J. and K. Lum. 1980. Patterns of plant species diversity in California: Relation to weather and topography. *American Naturalist* 116:504–536.

Rosen, D. E. 1978. Vicariant patterns and historical explanation in biogeography. *Systematic Zoology* 27:159–188.

Rosen, D. E. 1979. Fishes from the uplands and intermontane basins of Guatemala: Revisionary studies and comparative geography. *Bulletin of the American Museum of Natural History* 162:267–376.

Rosenzweig, M. 1975. On continental steady states of species diversity. In M. L. Cody and J. Diamond, eds., *Ecology and Evolution of Communities*, pp. 121–140. Cambridge, Mass.: Harvard University Press.

Rosenzweig, M. and J. A. Taylor. 1980. Speciation and diversity in Ordovician invertebrates: Filling niches quickly and carefully. Oikos 35:236–243.

Ross, H. H. 1972. The origin of species diversity in ecological communities. *Taxon* 21:253–259.

Schopf, T. J. M. 1974. Permo-Triassic extinctions: Relation to sea-floor spreading. *Journal of Geology* 82:129–143.

Sepkoski, J. J., Jr. 1978. A kinetic model of Phanerozoic taxonomic diversity. I. Analysis of marine orders. *Paleobiology* 4:223–251.

Sepkoski, J. J. Jr. 1979. A kinetic model of Phanerozoic taxonomic diversity. II. Early Phanerozoic families and multiple equilibria. *Paleobiology* 4:223–251

Sepkoski, J. J., Jr. 1981. A factor analytic description of the Phanerozoic marine fossil record. *Paleobiology* 7:36–53.

Sepkoski, J. J., Jr. 1984. A kinetic model of Phanerozoic taxonomic diversity. III. Post-Paleozoic families and mass extinctions. *Paleobiology* 10:246–267.

Sepkoski, J. J., Jr., R. K. Bambach, D. M. Raup, and J. W. Valentine. 1981. Phanerozoic marine diversity and the fossil record. *Nature* 293:435–437.

Shaffer, M. 1981. Minimum population sizes for species conservation. *Bioscience* 31:131–134.

Shaffer, M. 1987. Minimum viable populations: Coping with uncertainty. In M. E. Soulé, ed., *Viable Populations for Conservation*, pp. 69–86. Cambridge, U.K.: Cambridge University Press.

Signor, P. 1982. Species richness in the Phanerozoic: Compensating for sampling bias. *Geology* 10:625–628.

Simberloff, D. D. 1974. Permo-Triassic extinctions: Effects of an area on biotic equilibrium. *Journal of Geology* 82:267–274.

Simpson, G. G. 1964. Species density of North American Recent mammals. *Systematic Zoology* 13:57–73.

Soulé, M. E., ed. 1987. *Viable Populations for Conservation*. Cambridge, U.K.: Cambridge University Press.

Stanley, S. M. 1984. Marine mass extinctions: A dominant role for temperature. In M. Nitecki, ed., Extinctions, pp. 69–117. Chicago: University of Chicago Press.

Stanley, S. M. 1987. *Extinction.* New York: Scientific American Library.

Stanley, S. M. 1988. Paleozoic mass extinctions: Shared patterns suggest global cooling as a common cause. *American Journal of Science* 288:334–352.

Valentine, J. W. 1971. Plate tectonics and shallow marine diversity and endemism, an actualistic model. *Systematic Zoology* 20:253–264.

Valentine, J. W. 1972. Phanerozoic taxonomic diversity: A test of alternative models. *Science* 180:1078–1079.

Valentine, J. W. 1973. *Evolutionary Paleoecology of the Marine Biosphere.* Englewood Cliffs, N.J.: Prentice-Hall.

Valentine, J. W. 1977. General patterns of metazoan evolution. In A. Hallam, ed., *Patterns of Evolution, as Illustrated by the Fossil Record,* pp. 27–57. Amsterdam: Elsevier.

Valentine, J. W., ed. 1985. *Phanerozoic Diversity Patterns.* Princeton: Princeton University Press.

Valentine, J. W. and E. M. Moores. 1970. Plate-tectonic regulation of faunal diversity and sea level: A model. *Nature:* 228:657–659.

Valentine, J. W., and E. M. Moores. 1972. Global tectonics and the fossil record. *Journal of Geology* 80:167–184.

Van Valen, L. 1976. Energy and evolution. *Evolutionary Theory* 1:179–229.

Van Valen, L. and R. E. Sloan. 1977. Ecology and the extinction of dinosaurs. *Evolutionary Theory* 2:37–64.

Vermeij, G. J. 1989. Geographical restriction as a guide to the causes of extinction: The case of the cold northern oceans during the Neogene. *Paleobiology* 15: 335–356.

Vrba, E. S. 1985. Environment and evolution: Alternative causes of the temporal distribution of evolutionary events. *South African Journal of Science* 81:229–236.

Webb, S. D. 1969. Extinction-origination equilibria in Late Cenozoic land mammals of North America. *Evolution* 23:688–702.

Webb, S. D. 1976. Phanerozoic diversity patterns: Discussion. *Journal of Geology* 84:617–619.

West-Eberhard, M. J. 1983. Sexual selection, social competition, and speciation. *Quarterly Review of Biology* 58:155–183.

White, M. J. D. 1978. *Modes of Speciation.* San Francisco: W. H. Freeman.

Whittaker, R. H. 1972. Evolution and measurement of species diversity. *Taxon* 21:213–251.

Whittaker, R. H. 1977. Evolution of species diversity in land communities. *Evolutionary Biology* 10:1–67.

Wiens, J. A. 1989. Spatial scaling in ecology. *Functional Ecology* 3:385–397.

Wilson, A. C., G. L. Bush, S. M. Case, and M.-C. King. 1975. Social structuring of mammalian populations and rate of chromosomal evolution. *Proceedings of the National Academy of Sciences* 72:5061–5065.

6 : Phylogenetic and Ecologic Patterns in the Phanerozoic History of Marine Biodiversity

J. John Sepkoski, Jr.

Global biodiversity can be either fragile or robust, depending upon what temporal and spatial scales are examined. At small scales, diversity is often fragile: environmental perturbations and habitat destruction frequently decimate local species assemblages. At much larger scales, however, biodiversity appears more robust. Over the time scale of the Phanerozoic, diversity of multicellular organisms has increased greatly, despite constantly changing conditions on this dynamic planet. The net result of the last 600 myr of evolution is an uncertain diversity of somewhere between 5 million and 50 million species (May 1988).

Investigations of the fossil record at both local and global scales have indicated that the expansion of diversity did not proceed at continuous rates: there were long intervals spanning tens of million years of nearly constant global diversity, and there were abrupt intervals of massive decline in diversity—the mass extinctions that punctuate the fossil record. Rapid increases in diversity versus even more rapid declines reflect two different sets of phenomena:

1. Intrinsic abilities of different groups of animals and plants to diversify in local ecosystems, leading to expansion of their numbers globally

2. Extrinsic changes in our planet's environments, over time scales of

several million years down perhaps to several months or days, that
perturb ecosystems and cause loss of diversity

In this paper, I present a backdrop against which modern biodiversity can
be considered. I review some of what I think I know about the diversity history of multicellular organisms, concentrating on animals in the oceans. I begin with preliminary remarks on data from the fossil record and their limitations, and then discuss how global biodiversity has increased over the Phanerozoic. Next, I discuss the great declines in global diversity and how diversity
recovered from these. Finally, I briefly discuss relationships between global
diversity, which relates to phylogenetic patterns, and local diversity, which
relates to ecologic patterns.

Data on Fossil Diversity: The Necessity of Systematic Research

If one accepts that there are now around 50 million species on Earth, then the
scanty 1.5 million species that have been named (May 1988) represent but a
3% sample of modern biodiversity. There have probably been orders of magnitude more species that are now extinct, but the survey of the fossil record is
even poorer. As of 1970, only about 190,000 fossil species had been named
(Raup 1976). At least 95% of these were marine, and today the oceans hold
one to two orders of magnitude fewer species than terrestrial ecosystems do
(cf. Valentine 1970). Thus, the sample of past terrestrial ecosystems, especially the arthropod component, is very small indeed. If we restrict considerations to marine ecosystems, the survey of fossil species is still quite limited.
Estimates of the size of this sample have ranged from a fraction of a percent
to about 33% (Durham 1967; Signor 1985; Valentine 1970). I favor something
around 2%, derived from the following exercise:

1. The approximate number of animal species in the oceans today is
   somewhere around 200,000 (when endoparasites and gross ignorance are excluded).
2. The number of documented fossil marine animal species, with vertebrates added and increase accommodated since 1970, is probably
   about 250,000.
3. The first of these species appeared somewhere around 650 Ma.
4. The average duration of a marine animal species, integrated over the
   whole of the Phanerozoic, is probably around 5 myr.

These figures would be enough to calculate the total number of animal
species that ever lived in the oceans if we had a function describing what the

diversity of species through time looked like. We do not. But we can make a handy approximation if we assume that the diversity of higher taxa, namely, families, provides a good approximation of species diversity. Figure 6.1 illustrates the known history of familial diversity in the oceans. This history is irregular, but we can note that a straight-line approximation from 650 Ma to the present leaves about as much diversity below the line as above; that is, the errors cancel. Using this linear approximation, and scaling up to species, the calculation of the total number of marine animal species that ever lived becomes simply the area of the triangle measured in 5 myr increments:

$$0.5 \times 200,000 \text{ spp} \times 650 \text{ myr/5 myr} = 13,000,000 \text{ spp.} \quad (6.1)$$

Thus, the species sample represented by the known fossil record is

$$100 \times 250,000 \text{ spp} / 13,000,000 \text{ spp} \cong 2\%. \quad (6.2)$$

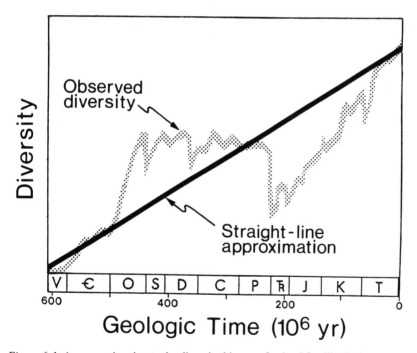

Figure 6.1 An approximation to the diversity history of animal families in the oceans. Although this history is irregular, a straight line from the appearance of marine animals around 650 Ma to the Tertiary maximum leaves about as much error above the line as below. Thus, the area under the straight line can be used to approximate the total number of marine animal taxa that ever lived.

Obviously, this is an order-of-magnitude, back-of-the-envelope type of calculation. Still, the result is small and would remain so even if scaled up by a factor of 2 to 3 for the proportion of skeletalized marine animals expected to leave a fossil record (Johnson 1964; Schopf 1978; Valentine 1970).

The small size of the sample underscores the need for continued systematic investigation of the biota, both living and extinct. What we know about the past record relied heavily on museum collections—the cumulation of centuries of investigation of the fossil record. The sample of past biodiversity will accumulate only with continued exploration of the fossil record (cf. recent advances in understanding the earliest Cambrian biota; see Cowie and Brasier 1989) and restudy of existing collections (cf. Culver, Buzas, and Collins 1987; Teichert, Sweet, and Boucot 1987).

A larger sample of the history of the Earth's biodiversity should be available in the systematic record of higher taxa: it takes but one species to establish the presence of a genus, family, etc. (Raup 1979a). Thus, these higher taxa should give us a more nearly complete picture of what has transpired over geologic time. There are problems, however, in that what constitutes a higher taxon has varied through the decades, and especially with new systematic philosophies. What was a genus to Linnaeus might be considered a family or even higher today (taxonomic inflation has existed), and what was a family to, say, G. G. Simpson might be considered an unnatural group to modern-day phylogenetic systematists.

This instability might constitute a serious problem if we wished to trace the history of a particular group through its superspecific taxonomy (cf. Patterson and Smith, 1987, 1989). Yet, I shall argue that for more encompassing groups—especially the whole-world fauna—this poses less of a problem. Consider an example outside the realm of biological systematics: the population of the United States. Most of this population is clustered in named municipalities of varying sizes, histories, and governmental structures. Many municipalities are parts of suburban sprawls where the end of one and beginning of another appears completely arbitrary, at least to someone raised in central New Jersey and now living in Chicago. Yet, the number of municipalities in a state provides a good estimate of the total population of that state. Figure 6.2 illustrates a plot of number of municipalities versus total state population. The correlation coefficient is 0.94 despite nearly four orders of magnitude variation in municipal sizes and arbitrary delineations of townlines in many suburban regions. This is true even though the town names, and their clustering into counties and states, give us little information about the history or causes of settlement.

This example is offered merely to establish that arbitrary groupings can still provide valuable information about underlying patterns, if the set of entities is sufficiently large (Sepkoski 1984). I have made similar arguments for ge-

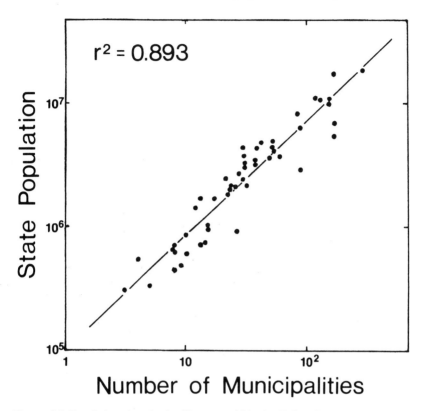

Figure 6.2 Population sizes in the 50 states within the United States versus number of municipalities with populations greater than 10,000 people in each. The strong correlation indicates that numbers of towns, even though variable in size and often arbitrary in demarcation, can be used to predict total state population. Data from the 1970 U.S. census.

nealogical systems, based on simulations of branching systems clustered into arbitrary paraclades (Sepkoski 1978, 1989, 1991a). More important, comparisons of real systematic entities have shown good correlations between species diversity and traditionally defined higher taxa in present-day situations (e.g., Jablonski and Flessa 1986) and over evolutionary time (e.g., Sepkoski et al. 1981; Bambach 1989). Maxwell and Benton (1990) have demonstrated that the picture of major features in tetrapod diversification, measured at the familial level, have not changed appreciably over the last 50 years, despite increases in the number of families recognized, changes in their known geologic ranges, and redefinitions within the recent cladistic framework. In my own data set on fossil families (Sepkoski 1982a), I have found few changes in detailed patterns of diversification despite changes in nearly half of the documented records accumulated over a decade of continued compilation.

Thus, we should be able to perceive something valid in the history of diversity from examining the fossil record of higher taxa. Still, the record we have is certainly sketchy—a world of Platonic shadows, if you will. Continued effort in systematics, and in collecting, will provide greater precision and resolution in our knowledge of biodiversity and its history at all taxonomic levels.

Phanerozoic Global Biodiversity: A History of Increase

It is trivial to note that biodiversity has increased from 3.5 Ga, the age of the oldest probable microbial fossils (Awramik, Schopf, and Walter 1983), to the present. It is less trivial to ask why multicellular organisms did not begin diversifying until perhaps 1.4 Ga (Walter, Du, and Horodyski 1990), nearly 60% of the way into the known history of life, and why multicellular animals did not diversify until around 0.6 Ga, more than 80% of the way into this history. Terrestrial plants and animals did not begin diversifying until around 0.45 Ga, almost 90% of the way into the history of life.

The basic observation here is that diversification has not been continuous. This is not a conclusion that could easily be reached from first evolutionary principles. Indeed, for animal diversification, hypothesized trajectories have ranged from nearly exponential increase (Cailleux 1954; Valentine 1970; Signor 1990) to an early maximum followed by decline (Raup 1972). The empirical history of animal diversity in the oceans falls between these extremes: there was no early maximum but neither was there continuous increase. Instead, the history was one of radiations and stabilizations, punctuated by mass extinctions.

Figure 6.3 shows that three phases of diversification can be recognized for marine animals at the family level:

1. Vendian-Cambrian phase, encompassing the earliest expansion in the latest Precambrian (i.e., Vendian Period), the explosive diversification during the Early Cambrian, and the stabilization during the Middle and Late Cambrian
2. later Paleozoic phase, initiated with the Ordovician radiations, which tripled familial (and generic) diversity (Sepkoski and Sheehan 1983; Sepkoski 1988), and followed by 200 million years of near steady state (despite several extinction events), terminated by the massive extinction at the end of the Permian Period
3. Mesozoic-Cenozoic phase, encompassing the rebound from the end-

Permian event and the continued expansion of diversity to a Neogene maximum

This qualitative pattern appears robust regardless of taxonomic and geographic scale examined (Sepkoski et al. 1981). Similar "stair-step" patterns, but with different timings, have been documented for terrestrial plants (Nik-

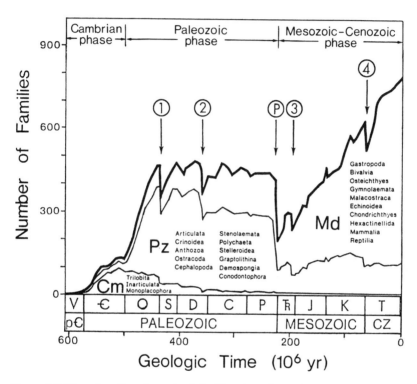

Figure 6.3 Diversity of marine animal families through the Phanerozoic. The upper curve shows the total number of well-skeletonized families known from each of 77 stratigraphic stages. This curve can be divided into three phases of diversification, as indicated at the top of the graph. The arrows above the curve identify major mass extinctions: P = end Permian, 1 = end Ordovician (Ashgillian), 2 = Late Devonian (Frasnian), 3 = end Triassic (Norian), and 4 = end Cretaceous (Maestrichtian). Fields below the upper curve delimit diversity histories of the three great evolutionary faunas: Cm = Cambrian fauna, Pz = Paleozoic fauna, and Md = Modern fauna; the principal classes in each fauna are listed. Geologic periods and eras are indicated along the abscissa: V = Vendian, Є = Cambrian, O = Ordovician, S = Silurian, D = Devonian, C = Carboniferous, P = Permian, Tr = Triassic, J = Jurassic, K = Cretaceous, and T = Tertiary; pЄ = Precambrian, and CZ = Cenozoic. Modified from Sepkoski (1991a).

las, Tiffney, and Knoll 1983; Knoll 1986), tetrapod vertebrates (Benton 1985, 1990), and, perhaps, even terrestrial arthropods (Sepkoski and Hulver 1985). These observations suggest that discontinuous increase in diversity is common to all components of the multicellular biosphere.

In the marine realm, the stair-step pattern of diversification was not a function of expansions of the entire fauna. Rather, it was affected by differential diversification among semidiscrete groups of higher taxa, which I have termed "evolutionary faunas" (Sepkoski 1981). Figure 6.3 illustrates the cumulative contributions of the three great evolutionary faunas to total diversity through the Phanerozoic. The initial diversification of animals resulted from expansion of the "Cambrian fauna," which included trilobites, inarticulate brachiopods, hyoliths, and stem groups of molluscs and echinoderms as well as various problematic taxa. These radiated rapidly early in the Cambrian Period, reached maximum diversity before its end, and declined thereafter. None participated to any extent in the Ordovician radiations. These radiations reflected expansion of the "Paleozoic fauna," which included articulate brachiopods, crinoids, corals, cephalopods, stenolaemate bryozoans, and other crown groups. Most of these originated during the Cambrian Period but did not undergo substantial radiation until the Early to Middle Ordovician. Collectively, they attained greatest diversity during the Devonian and declined afterward, slowly at first and then catastrophically at the end of the Permian. The final increase in marine biodiversity was affected by the "Modern fauna," which includes bivalve and gastropod molluscs, the various marine vertebrates, gymnolaemate bryozoans, malacostracan crustaceans, echinoids, and other classes. Most were present during the early Paleozoic but radiated only slowly. They survived the Permian mass extinction comparatively unscathed and then began a major diversification that pushed global marine biodiversity to unprecedented levels, beginning in the mid-Cretaceous Period.

This history suggests that the global expansion of diversity was not a function of continuous adaptation—or refinement—within established clades. Rather, it was caused by differential expansion of groups with unequal characteristic rates of taxic evolution that were more or less conserved through time (Sepkoski 1984; Van Valen 1985; Valentine 1990). For example, articulate brachiopods from the start had higher rates of origination and extinction than marine bivalves. Articulate brachiopods attained their greatest diversity during the Paleozoic Era, long before bivalves. But bivalves plodded on at low turnover rates through the Paleozoic Era and maintained these as they diversified into the Mesozoic and Cenozoic (Hallam and Miller 1988; Miller and Sepkoski 1989). At the same time, articulate brachiopods, with much reduced diversity, maintained higher rates of turnover (Valentine, Tiffney, and Sepkoski 1991). Thus, levels of biodiversity in the marine realm would seem more a function of the identity of the players than the act in the drama.

Why rates of taxic evolution are often conserved is not clear. Nor is why different taxa have different characteristic diversities, although there are some observations that bear upon this problem. Members of the Cambrian fauna appear mostly to have been ecological generalists with broad trophic and habitat requirements (Sepkoski 1979, 1988); their low beta diversity was certainly a component of their low global diversity. Members of the Paleozoic fauna, which included numerous sessile epifaunal suspension feeders, seemed to have had more specific trophic and habitat requirements and utilized greater amounts of ecospace (Bambach 1983) and habitat space (Ausich and Bottjer 1982; Bottjer and Ausich 1986). Broader utilization of resources and increased specialization thus appear to have increased local and global diversity (cf. "niche diversification hypothesis" of Connell 1978) and possibly to have caused displacement of the more generalist Cambrian fauna (cf. MacArthur 1972; Leigh 1990).

Ecological distinctions that would increase diversity between the Paleozoic and Modern evolutionary faunas are not so evident. The Modern fauna is characterized by a greater importance of predators (Vermeij 1977, 1983, 1987) and of infaunal bioturbators (Thayer 1979, 1983). These could have played roles in the displacement of the Paleozoic fauna, but their functions in diversity increase are not obvious. However, Bambach (1983) concluded that members of the Modern fauna utilize more ecospace than the earlier faunas, which is consistent with their greater within-community (Bambach 1977) and global diversity. It is also consistent with the idea that any exclusion of available ecospace, as by reengineering of the planet, can lead to extinction and decline in biodiversity over substantial time intervals.

Mass Extinctions and Their Aftermaths

The stair-step pattern of increasing diversity through the Phanerozoic was disturbed a number of times by extinction events of varying magnitudes and phylogenetic effects. Figure 6.4 illustrates the temporal distribution of recognized events and provides a rough indication of their relative magnitudes. Most of these events were devastating to some local ecosystems and eliminated on the order of 15% to 40% of marine species globally (Sepkoski 1989). But their phylogenetic effects were small, eliminating few families and even fewer recognized orders. Many of the events may have been regional, affecting limited geographic areas; examples are the Pliocene event of the North Atlantic (Stanley and Campbell 1981; Stanley 1986), Tithonian and lower Toarcian events of Europe and North America (Hallam 1986), and, perhaps, Cambrian "biomere" events of North America and sometimes China and Australia (Palmer 1979, 1982).

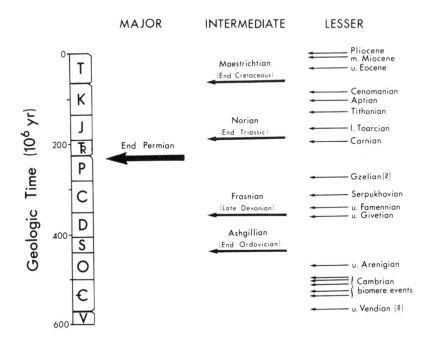

Figure 6.4 The Phanerozoic distribution through time of known extinction events, classified into major, intermediate, and lesser based on magnitudes of extinction and diversity loss. Abbreviations for geologic periods are as in figure 6.3. Updated from Sepkoski (1982b).

Only five events—those properly termed "mass extinctions" (Flessa et al. 1986)—were of sufficient magnitude and global expanse to have profound phylogenetic effects (figure 6.3). Four of these are estimated to have eliminated 65% to 75% of animal species in the oceans (cf. Sepkoski 1989): the end-Ordovician, late Devonian, end-Triassic, and end-Cretaceous ("K/T") mass extinctions. The single great big one, that at the end of the Permian, may have excised more than 95% of marine species (Raup 1979b) and affected the transition from a world ocean dominated by the Paleozoic evolutionary fauna to an ocean dominated by the Modern fauna.

The nature of the perturbations that cause mass extinction, the time scales over which they operate, the sequences of ecosystem collapse, and the characteristics of victims and survivors are all currently topics of extensive research and debate (e.g., Stanley 1987; Larwood 1988; Donovan 1989; Kauffman and Walliser 1990; Sharpton and Ward 1991). Less attention has been focused on events immediately following these crises and on the evolutionary recovery of biodiversity.

The larger extinction events appear to have been followed by extended intervals of low diversity, lasting from several tens of thousand years as in the case of the end-Cretaceous (Smit 1982; Keller 1988) to several million years in the case of the end-Permian (Newell 1967; Kummel 1973). During these intervals, various holdover species from before the crises disappeared; examples include some Cretaceous foraminifera during the earliest Danian (Keller 1988), some relict Ordovician assemblages during the earliest Silurian (Baarli 1987), and some Permian brachiopods in what is probably earliest Triassic (Sheng et al. 1984). At the ends of these intervals, marine faunas were characterized by large numbers of cosmopolitan taxa (i.e., low gamma diversities), probably as a result of loss of endemics (Bretsky 1973; Boucot 1975; Sheehan 1982, 1988; Jablonski 1986a,b; Erwin 1989). The few paleoecologic studies that have been conducted on these faunas, such as those of earliest Silurian assemblages (Sheehan 1975, 1982, 1988), suggest that community structure was characterized by reduced numbers of species in local assemblages (i.e., low alpha diversities) and by a few assemblages spread widely over environmental gradients (i.e., low beta diversities) (see also Kummel 1973; Boucot 1983; Sheehan 1985; Hansen 1988; Stanley 1990). In a nutshell, these large crises hit biodiversity at all its components.

The postcrisis intervals were followed by recovery of diversity and return to the "normal" state of the ecosystem. Diversity rebounded either to previous levels, as seen during the Paleozoic Era, or to the long-term trend of expansion, as seen during the Mesozoic and Cenozoic (figure 6.3). The nature of the diversity rebounds is probably best exemplified by the Silurian recovery from the end-Ordovician mass extinction, illustrated in figure 6.5A. This crisis appears to have been the second largest mass extinction in history, eliminating 27% of marine animal families and nearly 50% of marine genera. During the Early Silurian, diversification rates were comparable to those of the Ordovician radiations. These rates tapered as diversity approached preevent levels, and rebound finally ceased in the Middle Silurian when diversity reached the level of the Late Ordovician, where it remained, more or less, into the Devonian (figure 6.3). Thus, the recovery of global biodiversity required on the order of 12 to 15 million years.

Rapid rebound to preevent state is probably the most compelling evidence that global diversity behaves as an equilibrium (or quasi-equilibrium) system. Expectations for such a system shocked by episodic perturbations have been modeled by Carr and Kitchell (1980), Sepkoski (1984), Kitchell and Carr (1985), Valentine and Walker (1987), McKinney (1989), and Miller and Sepkoski (1989) (see also Stanley 1979). The general structure for a homogeneous equilibrial system (i.e., one with a single evolutionary fauna) is illustrated in figure 6.5B. Speciation and extinction rates are depicted as functions of standing diversity (for justification, see MacArthur 1969; Rosenzweig

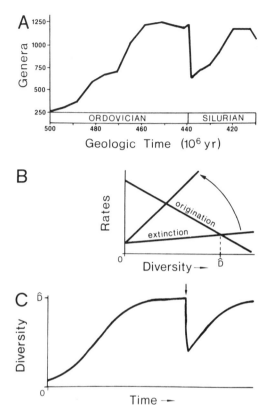

Figure 6.5 Observed and modeled patterns of rebound after mass extinctions. A. Genus-level diversity for marine animals through the Ordovician radiations, end-Ordovician mass extinction, and Silurian rebound. Diversity begins recovering soon after the mass extinction, and recovers at about the same rate as the preceding radiations; recovery ends when diversity reaches approximately its previous level. B. The elements of a simple, logistic model of diversification. The per-taxon rate (or probability) of origination is treated as a steeply declining linear function of diversity, whereas extinction is treated as a mildly increasing function; the two functions cross at the equilibrium diversity, $\hat{D}$. An "extinction event," or perturbation, can be generated by temporarily increasing the extinction function. C. A solution to the logistic model with a perturbation and rebound. Diversity increases sigmoidally from some starting value, D_O, and then levels at its equilibrium. The perturbation drives diversity downward. When the original extinction function is restored, diversity rebounds with the same trajectory as in the initial diversification and levels at the previous equilibrium. Figure 6.5A compiled from data described in Sepkoski (1986); figure 6.5B and C after Sepkoski (1984).

1975; Sepkoski 1978, 1991a; Maurer 1989), and a perturbation is treated as a temporary steepening of the extinction function (or it could be treated as a temporary increase in the *y*-axis intercept of the extinction function).

In such a system, diversity grows logistically, similarly to what is observed across the Precambrian-Cambrian Boundary and during the Ordovician radiations (Sepkoski 1984): there is approximately exponential growth when diversity is low, decelerating growth as diversity approaches the equilibrial carrying capacity, and finally steady-state maintenance when diversity is at the equilibrium and the system is undisturbed (figure 6.5C). Increase in extinction rate during a perturbation forces diversity below the carrying capacity, affecting a mass extinction; diversity decays, however, down to the new effective equilibrium level, so that diversity loss continues after the main pulse of extinction. When extinction is returned to its normal function, diversity rebounds, following the same trajectory as during the initial, predisturbance radiation; thus, the duration and total shape of the rebound depend upon how low diversity was depressed, with long lag phases of diversification possible if diversity drops sufficiently low (Carr and Kitchell 1980). After attainment of the original carrying capacity, diversity remains in steady state until the next perturbation. This is, of course, the pattern seen in the Silurian rebound.

There are other patterns observed during rebounds that are not encompassed by such simple models:

1. During the early stages of recovery, there can be rapid turnover of species and higher taxa, suggesting some kind of evolutionary instability before the system settles down and species of normal longevity reappear; examples include short-lived trilobite species after some biomere events (Stitt 1971, 1975; Hardy 1985), rapid community replacement during the earliest Silurian (Sheehan 1975), very rapid turnover of Scythian ammonoids after the end-Permian mass extinction (see House and Senior 1981), and boom-and-bust patterns among planktonic forminifera and benthic molluscs after the end-Cretaceous event (Smit 1982; Smit and Romein 1985; Hansen 1988; Keller 1988).

2. Community-level recoveries may occur at a variety of spatial scales, including increase in numbers of species within communities, increase in differentiation between communities, and increase in endemicity within broad regions, such as seen during the Silurian recovery (Cocks and McKerrow 1978; Sheehan 1975, 1980, 1982, 1988).

3. Unusual numbers of major evolutionary novelties (i.e., species with unique morphologies) may appear during rebounds. Erwin, Valentine, and Sepkoski (1987) provide data indicating that rebounds after

the Phanerozoic mass extinctions produced new taxonomic orders (presumed to be surrogates for major novelties) at a rate nearly twice that of normal, "background" times (see also Sheehan 1982; Jablonski 1986b). This tendency might have had something to do with ecological opportunities provided in depauperate ecosystems (nearly empty "ecological barrels"), although it might simply reflect the large numbers of species produced during rebounds.

Global, Phylogenetic Diversity and Local, Ecologic Diversity

Eldredge (1985, 1989, this volume) has argued that there is an important ontological difference between phylogenetic diversity—the number of species, genera, etc. within a clade—and ecologic diversity—the number of species, etc. at a given place. The former is a function of genealogical processes: speciation resulting from reproductive isolation of populations and extinction resulting from death of all members of a gene pool (formerly given cohesion by interbreeding and gene flow). Ecologic diversity is a function of "economic" processes: the number of individuals supported by resources in a given area and the manner in which these individuals are apportioned among species of differing trophic needs, habitat requirements, etc.

Although phylogenetic diversity and ecological diversity may be distinct quantities, there is reason to believe they have been closely associated over evolutionary time. The remarks above on mass extinction emphasized that as these events reduced global phylogenetic diversity of species within clades, they also lowered ecologic diversity by reducing species numbers within habitats and faunal differentiation between habitats. Rebounds from mass extinctions involved restoration of global diversity as well as reestablishment of local species numbers and community differentiation. This need not have been the case: obliteration of Australia would cause a decline in global mammalian diversity, or of Madagascar, a decline in global primate diversity (now tragically occurring), but neither would affect local communities over the rest of the world.

Strong association between local and global diversity is also seen in the great expansions of marine biodiversity. Bambach (1977) compiled data showing that numbers of species within marine benthic communities through the Phanerozoic paralleled, in a coarse way, the global diversity of marine animals. More recently (Sepkoski 1988), I showed that even though correlated, local and global diversity exhibited some quantitative discrepancies during the Paleozoic Era; these could be accounted for in changes in beta diversity—differentiation between local communities. Thus, the degree of habitat selection among species—an ecological parameter relating to resource partitioning—had important influence on global diversity during large radiations.

This parallel extends to taxonomic components of the global fauna. Figure 6.6 illustrates patterns of local, genus-level diversity on the marine shelf for selected higher taxa during the Paleozoic Era and compares these to their patterns of global diversity. These diagrams were constructed from 505 published faunal lists from North America assembled by A. I. Miller and myself (initially reported in Sepkoski and Miller 1985). These lists were arrayed into six generalized environmental zones, running from nearshore, peritidal habitats (zone 1), to nearshore protected environments (zone 2), offshore shoaling areas (zone 3), open midshelf environments (zone 4), outer-shelf environments (zone 5), and finally off-shelf and basinal areas (zone 6) (for more detail, see Sepkoski 1987, 1988; Miller 1988; also Bottjer and Jablonski 1988; Jablonski and Bottjer 1990a,b). The contoured time-environment diagrams were constructed by averaging the number of genera within each taxon over all faunal lists in a given environmental zone and stratigraphic series and then contouring these means at intervals of one genus (see also Miller 1988, 1990; Sepkoski 1991a,b).

All four illustrated taxonomic groups exhibit correlation between local, ecologic diversity and global, phylogenetic diversity in both their radiations and their declines. The Trilobita (of the Cambrian evolutionary fauna) attained maximum global diversity in the late Middle and early Late Cambrian, a time when it exhibited its maximum alpha diversity on the North American shelf. Later, during the Ordovician Period, trilobites lost diversity on the shelf, beginning in nearshore habitats and extending later to the offshore; this loss was paralleled in declining global diversity. In the later part of the Paleozoic Era, trilobites were minor components of local communities, and their global diversity was accordingly low. (The exception occurred during the Devonian Period when local diversity was declining but global diversity exhibited a minor expansion; this evidently reflected a regional radiation in the "Old World Province" that did not reach North America [Chlupàc 1975].)

The other taxonomic groups are similar in relationship if dissimilar in history. Strophomenid brachiopods, members of the Paleozoic evolutionary fauna, radiated in mid to outer shelf environments during the Ordovician Period and remained important there until the end of the Permian. Each of the maxima in their global diversity, separated by extinction events of varying magnitude, was reflected in environmentally circumscribed maxima of local diversity on the shelf. Rugose corals experienced a somewhat similar history of global diversity and occupation of local environments with maxima, and especially minima, in diversity reflected both globally and locally. The pterioid bivalves, members of the Modern fauna, underwent a somewhat different history of more nearly continuous expansion of local diversity and habitat

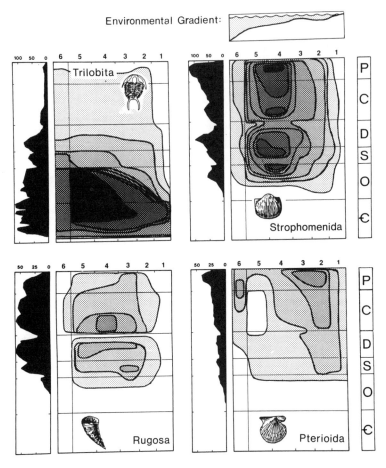

Figure 6.6 Time-environment diagrams for four taxonomic groups illustrating patterns of local genus-level diversity on the shelf through the Paleozoic Era and comparing these to patterns of global genus-level diversity (narrow graphs). Contours on the diagrams are at intervals of one genus, starting at zero, and stippling darkens toward higher diversities. Data on global genus-level diversity are from the compilation described in Sepkoski (1986). In all four groups, there is a qualitative association between highs and lows of local, ecologic diversity and of global, phylogenetic diversity. From Sepkoski and Miller (in preparation).

occupation; this again was reflected in more nearly continuously increasing global diversity through the Paleozoic Era (which began somewhat before pterioids became common on the North American shelf). Miller (1988) provides other examples from the Bivalvia during the Paleozoic Era, and Bottjer and Jablonski (1988) and Jablonski and Bottjer (1990a,b) present examples from the Mesozoic and Cenozoic, proving that the association between local and global diversities was not confined to any one interval of Phanerozoic time.

The empirical association of phylogenetic and ecologic diversity could be a function of either of two phenomena (which need not be mutually exclusive):

1. Phylogenetic success could provide a pool from which local communities are stocked.
2. Success in local communities could sum to large phylogenetic diversity.

I suspect the exceptions to the association—such as the Devonian radiation of European trilobites—favor the second explanation, that phylogenetic diversity is a function of local, ecological processes.

On the basis of this conclusion, I have investigated a simple model for taxic replacement in marine environments, with emphasis on the onshore-offshore shifts in centers of ecologic diversity evident for several taxa illustrated in figure 6.6 (i.e., trilobites and pterioids). An important finding in this work is that the mathematical structure needed to model local changes in diversity and faunal dominance is the same as that developed to describe global, phylogenetic diversity (cf. figure 6.5B). The details of this exercise are presented in Sepkoski (1991b).

I have argued three principal points in this essay:

1. Biodiversity has tended to increase over Phanerozoic time, but this increase is very much dependent upon the kinds of animals dominating the global fauna.
2. The past perturbations that have caused massive loss of diversity in marine ecosystems have been followed by rebounds to the level present prior to the events but often with novel taxa.
3. Much of what we measure in the global, phylogenetic diversity of the world-ocean fauna or of individual clades is a reflection of what was happening in local ecosystems, integrated over the entire globe.

These conclusions can provide either concern or solace, depending upon one's point of view. If phylogenetic success is dependent upon local ecological success, then human disturbance is going to alter the course of evolution irreparably. On the other hand, the biota should recover from human interference and might even recover with evolutionary novelty, *given that this disturbance ends*. But the bad news still is that the time scale for global recovery will be long relative to human standards—either the experience of an individual or the history of a culture. Consider that mammalian faunas have not yet begun to recover from the decimation 10,000 years ago (Martin and Klein 1984; Janis and Damuth 1990), and we live in a world without a normal complement of large terrestrial animals for our fascination and utilization.

The dynamic considerations summarized here are sketchy. More detail is presented in many of the references cited, but these, too, are very incomplete because of the limits of scientific insight and because of the imperfections of the data that inspire and test ideas. The former can be rectified with time, talent, and training. The latter needs concerted effort to describe the Earth's biota, both present and past. Biological systematics, the forte of museums and their personnel, plays an essential role in providing new data on biotic history and in clarifying the data already at hand. Improved data are absolutely necessary so that the dynamic processes of diversification and the constraints on fragility and recovery can be more readily rationalized—and, as necessary, engineered—by the world community as a whole.

ACKNOWLEDGMENTS

A. I. Miller collaborated in the research behind figure 6.6. Research for this essay received partial support from NASA grant NAGW-1693.

REFERENCES

Ausich, W. I. and D. J. Bottjer. 1982. Tiering in suspension-feeding communities in soft substrata through the Phanerozoic. *Science* 216:173–174.

Awramik, S. M., J. W. Schopf, and M. R. Walter, 1983. Filamentous fossil bacteria from the Archean of Western Australia. *Precambrian Research* 20:357–374.

Baarli, B. G. 1987. Benthic faunal associations in the Lower Silurian Solvik Formation of the Oslo-Asker District, Norway. *Lethaia* 20:75–90.

Bambach, R. K. 1977. Species richness in marine benthic habitats through the Phanerozoic. *Paleobiology* 3:152–167.

Bambach, R. K. 1983. Ecospace utilization and guilds in marine communities through the Phanerozoic. In M. J. S. Tevesz and P. L. McCall, eds., *Biotic Interactions in Recent and Fossil Benthic Communities*, pp. 719–746. New York: Plenum.

Bambach, R. K. 1989. Similarities and differences in diversity patterns at different taxonomic levels using traditional (non-cladistic) groupings. *Geological Society of America Abstracts with Program* 21(6):A206-A207.

Benton, M. J. 1985. Patterns in the diversification of Mesozoic non-marine tetrapods and problems in historical diversity analysis. *Special Papers in Palaeontology* 33:185–202.

Benton, M. J. 1990. Reptiles. In K. J. McNamara, ed., *Evolutionary Trends*, pp. 279–300. London: Belhaven Press.

Bottjer, D. J. and W. I. Ausich. 1986. Phanerozoic development of tiering in soft substrata suspension-feeding communities. *Paleobiology* 12:400–420.

Bottjer, D. J. and D. Jablonski. 1988. Paleoenvironmental patterns in the evolution of post-Paleozoic benthic marine invertebrates. *Palaios* 3:540–560.

Boucot, A. J. 1975. *Evolution and Extinction Rate Controls.* Amsterdam: Elsevier.

Boucot, A. J. 1983. Does evolution take place in an ecological vacuum? II. *Journal of Paleontology* 57:1–30.

Bretsky, P. W. 1973. Evolutionary patterns in the Paleozoic Bivalvia: Documentation and some theoretical considerations. *Geological Society of America Bulletin* 84:2079–2096.

Cailleux, A. 1954. How many species? *Evolution* 8:83–84.

Carr, T. R. and J. A. Kitchell. 1980. Dynamics of taxonomic diversity. *Paleobiology* 6:427–443.

Chlupàc, I. 1975. The distribution of phacopid trilobites in space and time. *Fossils and Strata* 4:399–408.

Cocks, L. R. M. and W. S. McKerrow. 1978. Silurian. In W. S. McKerrow, ed., *The Ecology of Fossils,* pp. 93–124. Cambridge, Mass.: MIT Press.

Connell, J. H. 1978. Diversity in tropical rain forests and coral reefs. *Science* 199:1302–1310.

Cowie, J. W. and M. D. Brasier, eds. 1989. *The Precambrian-Cambrian Boundary.* Oxford: Clarendon Press.

Culver, S. J., M. A. Buzas, and L. S. Collins. 1987. On the value of taxonomic standardization in evolutionary studies. *Paleobiology* 13:169–176.

Donovan, S. K., ed. 1989. *Mass Extinctions: Processes and Evidence.* New York: Columbia University Press.

Durham, J. W. 1967. The incompleteness of our knowledge of the fossil record. *Journal of Paleontology* 41:559–565.

Eldredge, N. 1985. *Unfinished Synthesis: Biological Hierarchies and Modern Evolutionary Thought.* New York: Oxford University Press.

Eldredge, N. 1989. *Macroevolutionary Dynamics: Species, Niches and Adaptive Peaks.* New York: McGraw-Hill.

Erwin, D. H. 1989. The end-Permian mass extinction: What really happened and did it matter? *Trends in Ecology and Evolution* 4:225–229.

Erwin, D. H., J. W. Valentine, and J. J. Sepkoski, Jr. 1987. A comparative study of diversification events: The early Paleozoic versus the Mesozoic. *Evolution* 41:1177–1186.

Flessa, K. W., H. K. Erben, A. Hallam, K. J. Hsü, H. M. Hüssner, D. Jablonski, D. M. Raup, J. J. Sepkoski, Jr., M. E. Soulé, W. Stinnesbeck, and G. J. Vermeij. 1986. Causes and consequences of extinction. In D. M. Raup and D. Jablonski, eds., *Patterns and Process in the History of Life,* pp. 235–257. Berlin: Springer-Verlag.

Hallam, A. 1986. The Pliensbachian and Tithonian extinction events. *Nature* 319:765–768.

Hallam, A. and A. I. Miller. 1988. Extinction and survival in the Bivalvia. In G. P. Larwood, ed., *Extinction and Survival in the Fossil Record,* pp. 121–138. Oxford: Clarendon Press.

Hansen, T. A. 1988. Early Tertiary radiation of marine molluscs and the long-term effects of the Cretaceous-Tertiary extinction. *Paleobiology* 14:37–51.

Hardy, M. C. 1985. Testing for adaptive radiation: The Ptychaspid (Trilobita) Biomere of the Late Cambrian. In J. W. Valentine, ed., *Phanerozoic Diversity Patterns: Profiles in Macroevolution,* pp. 379–399. Princeton, N.J.: Princeton University Press.

House, M. R. and J. R. Senior, eds. 1981. *The Ammonoidea.* Systematics Association special volume 18. London: Academic Press.

Jablonski, D. 1986a. Background and mass extinctions: The alternation of macroevolutionary regimes. *Science* 231:129–133.

Jablonski, D. 1986b. Evolutionary consequences of mass extinctions. In D. M. Raup and D. Jablonski, eds., *Patterns and Processes in the History of Life,* pp. 313–329. Berlin: Springer-Verlag.

Jablonski, D. and D. J. Bottjer. 1990a. Onshore-offshore trends in marine invertebrate evolution. In R. M. Ross and W. D. Allmon, eds., *Causes of Evolution: A Paleontological Perspective,* pp. 21–75. Chicago: University of Chicago Press.

Jablonski, D. and D. J. Bottjer. 1990b. The origin and diversification of major groups: environmental patterns and macroevolutionary lags. In P. D. Taylor and G. P. Larwood, eds., *Major Evolutionary Radiations,* pp. 17–57. Systematics Association special volume 42. Oxford: Clarendon Press.

Jablonski, D. and K. W. Flessa. 1986. The taxonomic structure of shallow-water marine faunas: Implications for Phanerozoic extinctions. *Malacologia* 27:43–66.

Janis, C. M. and J. Damuth. 1990. Mammals. In K. J. McNamara, ed., *Evolutionary Trends,* pp. 301–345. London: Belhaven Press.

Johnson, R. G. 1964. The community approach to paleoecology. In J. Imbrie and N. Newell, eds., *Approaches to Paleoecology,* pp. 107–134. New York: Wiley.

Kauffman, E. G. and O. H. Walliser, eds. 1990. *Extinction Events in Earth History.* Berlin: Springer-Verlag.

Keller, G. 1988. Extinction, survivorship and evolution of planktonic foraminifera across the Cretaceous/Tertiary Boundary at El Kef, Tunisia. *Marine Micropaleontology* 13:239–263.

Kitchell, J. A. and T. R. Carr. 1985. Nonequilibrium models of diversification: Faunal turnover dynamics. In J. W. Valentine, ed., *Phanerozoic Diversity Patterns: Profiles in Macroevolution,* Princeton: Princeton University Press.

Knoll, A. H. 1986. Patterns of change in plant communities through geological time. In J. Diamond and T. J. Case, eds., *Community Ecology,* pp. 126–141. New York: Harper and Row.

Kummel, B. 1973. Lower Triassic (Scythian) molluscs. In A. Hallam, ed., *Atlas of Palaeobiogeography,* pp. 225–233. Amsterdam: Elsevier.

Larwood, G. P., ed. 1988. *Extinction and Survival in the Fossil Record.* Oxford: Clarendon Press.

Leigh, E. G., Jr. 1990. Community diversity and environmental stability: A reexamination. *Trends in Ecology and Evolution* 5:340–344.

MacArthur, R. H. 1969. Patterns of communities in the tropics. *Biological Journal of the Linnaean Society* 1:19–30.

MacArthur, R. H. 1972. *Geographical Ecology.* New York: Harper and Row.

Martin, P. S. and R. G. Klein, eds. 1984. *Quaternary Extinctions*. Tucson: University of Arizona Press.

Maurer, B. A. 1989. Diversity-dependent species dynamics: Incorporating effects of population-level processes on species dynamics. *Paleobiology* 15:133–146.

Maxwell, W. D. and M. J. Benton. 1990. Historical tests of the absolute completeness of the fossil record of tetrapods. *Paleobiology* 16:322–335.

May, R. M. 1988. How many species are there on Earth? *Science* 241:1441–1449.

McKinney, M. L. 1989. Periodic mass extinctions: Product of biosphere growth dynamics? *Historical Biology* 2:273–287.

Miller, A. I. 1988. Spatio-temporal transitions in Paleozoic Bivalvia: An analysis of North American fossil assemblages. *Historical Biology* 1:251–273.

Miller, A. I. 1990. Bivalves. In K. J. McNamara, ed., *Evolutionary Trends*, London: Belhaven Press.

Miller, A. I. and J. J. Sepkoski, Jr. 1989. Modeling bivalve diversification: The effect of interaction on a macroevolutionary system. *Paleobiology* 14:364–369.

Newell, N. D. 1967. Revolutions in the history of life. *Geological Society of America Special Paper* 89:63–91.

Niklas, K. J., B. H. Tiffney, and A. H. Knoll. 1983. Patterns in vascular land plant diversification. *Nature* 303:614–616.

Palmer, A. R. 1979. Biomere boundaries re-examined. *Alcheringa* 3:33–41.

Palmer, A. R. 1982. Biomere boundaries: A possible test for extraterrestrial perturbation of the biosphere. *Geological Society of American Special Paper* 190:469–475.

Patterson, C. and A. B. Smith. 1987. Is the periodicity of extinctions a taxonomic artifact? *Nature* 330:248–251.

Patterson, C. and A. B. Smith. 1989. Periodicity in extinction; The role of systematics. *Ecology* 70:802–811.

Raup, D. M. 1972. Taxonomic diversity during the Phanerozoic. *Science* 215:1065–1071.

Raup, D. M. 1976. Species diversity in the Phanerozoic: A tabulation. *Paleobiology* 2:279–288.

Raup, D. M. 1979a. Biases in the fossil record of species and genera. *Carnegie Museum of Natural History Bulletin* 13:85–91.

Raup, D. M. 1979b. Size of the Permo-Triassic bottleneck and its evolutionary implications. *Science* 206:217–218.

Rosenzweig, M. L. 1975. On continental steady states of species diversity. In M. L. Cody and J. M. Diamond, eds., *Ecology and Evolution of Communities*, pp. 121–140. Cambridge, Mass.: Belknap Press.

Schopf, T. J. M. 1978. Fossilization potential of an intertidal fauna: Friday Harbor, Washington. *Paleobiology* 4:261–270.

Sepkoski, J. J., Jr. 1978. A kinetic model of Phanerozoic taxonomic diversity. I. Analysis of marine orders. *Paleobiology* 4:223–251.

Sepkoski, J. J., Jr. 1979. A kinetic model of Phanerozoic taxonomic diversity. II. Early Phanerozoic families and multiple equilibria. *Paleobiology* 5:222–251.

Sepkoski, J. J., Jr. 1981. A factor analytic description of the Phanerozoic marine fossil record. *Paleobiology* 7:36–53.

Sepkoski, J. J., Jr. 1982a. A compendium of fossil marine families. *Milwaukee Public Museum Contributions in Biology and Geology*, no. 51.

Sepkoski, J. J., Jr. 1982b. Mass extinction in the Phanerozoic oceans: A review. *Geological Society of America Special Paper* 190:283–289.

Sepkoski, J. J., Jr. 1984. A kinetic model of Phanerozoic taxonomic diversity. III. Post-Paleozoic families and mass extinctions. *Paleobiology* 10:246–267.

Sepkoski, J. J., Jr. 1986. Phanerozoic overview of mass extinction. In D. M. Raup and D. Jablonski, eds., *Patterns and Processes in the History of Life*, pp. 277–295. Berlin: Springer-Verlag.

Sepkoski, J. J., Jr. 1987. Environmental trends in extinction during the Paleozoic. *Science* 235:64–66.

Sepkoski, J. J., Jr. 1988. Alpha, beta, or gamma: Where does all the diversity go? *Paleobiology* 14:221–234.

Sepkoski, J. J., Jr. 1989. Periodicity in extinction and the problem of catastrophism in the history of life. *Journal of the Geological Society of London* 146:7–19.

Sepkoski, J. J., Jr. 1991a. Diversity in the Phanerozoic oceans: A partisan review. In Dudley, E., ed., *Fourth International Congress of Systematic and Evolutionary Biology, Proceedings*. Portland, Ore.: Dioscorides Press.

Sepkoski, J. J., Jr. 1991b. A model of onshore-offshore change in faunal diversity. *Paleobiology* 17:58–77.

Sepkoski, J. J., Jr., R. K. Bambach, D. M. Raup, and J. W. Valentine. 1981. Phanerozoic marine diversity and the fossil record. *Nature* 293:435–437.

Sepkoski, J. J., Jr. and M. L. Hulver. 1985. An atlas of Phanerozoic clade diversity diagrams. In J. W. Valentine, ed., *Phanerozoic Diversity Patterns: Profiles in Macroevolution*, pp. 11–39. Princeton: Princeton University Press.

Sepkoski, J. J., Jr. and A. I. Miller. 1985. Evolutionary faunas and the distribution of Paleozoic marine communities in space and time. In J. W. Valentine, ed., *Phanerozoic Diversity Patterns: Profiles in Macroevolution*, pp. 153–190. Princeton: Princeton University Press.

Sepkoski, J. J., Jr., and P. M. Sheehan. 1983. Diversification, faunal change, and community replacement during the Ordovician radiations. In M. J. S. Tevesz and P. L. McCall, eds., *Biotic Interactions in Recent and Fossil Benthic Communities*, pp. 673–717. New York: Plenum.

Sharpton, V. L. and P. D. Ward, eds. 1991. Global catastrophes in earth history. *Geological Society of America Special Paper* 247.

Sheehan, P. M. 1975. Brachiopod synecology in a time of crisis (late Ordovician-Early Silurian). *Paleobiology* 1:205–212.

Sheehan, P. M. 1980. Paleogeography and marine communities of the Silurian carbonate shelf in Utah and Nevada. In T. D. Fouch and E. R. Megathan, eds., *Paleozoic Paleogeography of West-Central United States*, pp. 19–37. Rocky Mountain Section, Society of Economic Paleontologists and Mineralogists.

Sheehan, P. M. 1982. Brachiopod macroevolution at the Ordovician-Silurian Boundary. *Third North American Paleontological Convention, Proceedings* 2:477–481.

Sheehan, P. M. 1985. Reefs are not so different—they follow the evolutionary pattern of level-bottom communities. *Geology* 13:46–49.

Sheehan, P. M. 1988. Late Ordovician events and the terminal Ordovician extinction. *New Mexico Bureau of Mines and Mineral Resources Memoir* 44:405–415.

Sheng J.-Z., Chen, C.-Z., Wang, Y.-G., Rui, L., Liao, Z.-T., Y. Bando, Ishii, K.-I., K. Nakazawa, and K. Nakamura. 1984. Permian-Triassic Boundary in middle and eastern Tethys. *Journal of the Faculty of Sciences, Hokkaido University,* ser. IV, 21:133–181.

Signor, P. W., III. 1985. Real and apparent trends in species richness through time. In J. W. Valentine, ed., *Phanerozoic Diversity Patterns: Profiles in Macroevolution,* pp. 129–150. Princeton: Princeton University Press.

Signor, P. W., III. 1990. The geologic history of diversity. *Annual Review of Ecology and Systematics* 21:509–539.

Smit, J. 1982. Extinction and evolution of planktonic foraminifera after a major impact at the Cretaceous/Tertiary Boundary. *Geological Society of America Special Paper* 190:329–352.

Smit, J. and A. J. T. Romein. 1985. A sequence of events across the Cretaceous-Tertiary Boundary. *Earth and Planetary Science Letters* 74:155–170.

Stanley, S. M. 1979. *Macroevolution: Pattern and Process.* San Francisco: W. H. Freeman.

Stanley, S. M. 1986. Anatomy of a regional mass extinction: Plio-Pleistocene decimation of the Western Atlantic bivalve fauna. *Palaios* 1:17–36.

Stanley, S. M. 1987. *Extinction.* New York: Scientific American Books.

Stanley, S. M. 1990. Delayed recovery and the spacing of major extinctions. *Paleobiology* 16:401–414.

Stanley, S. M. and L. D. Campbell. 1981. Neogene mass extinction of Western Atlantic molluscs. *Nature* 293:457–459.

Stitt, J. H. 1971. Repeating evolutionary pattern in Late Cambrian trilobite biomeres. *Journal of Paleontology* 45:178–181.

Stitt, J. H. 1975. Adaptive radiation, trilobite paleoecology and extinction, Ptychaspid Biomere, Late Cambrian of Oklahoma. *Fossils and Strata* 4:381–390.

Teichert, C., W. C. Sweet, and A. J. Boucot. 1987. The unpublished fossil record: Implications. *Senckenbergiana Lethaea* 68:1–19.

Thayer, C. W. 1979. Biological bulldozers and the evolution of marine benthic communities. *Science* 203:458–461.

Thayer, C. W. 1983. Sediment-mediated biological disturbance and the evolution of marine benthos. In M. J. S. Tevesz and P. L McCall, eds., *Biotic Interactions in Recent and Fossil Benthic Communities,* pp. 480–625. New York: Plenum.

Valentine, J. W. 1970. How many marine invertebrate species? A new approximation. *Journal of Paleontology* 44:410–415.

Valentine, J. W. 1990. The macroevolution of clade shape. In R. M. Ross and W. D. Allmon, eds., *Causes of Evolution: A Paleontological Perspective,* pp. 128–150. Chicago: University of Chicago Press.

Valentine, J. W., B. H. Tiffney, and J. J. Sepkoski, Jr. 1991. Evolutionary dynamics of plants and animals: A comparative approach. *Palaios* 6:81–88.

Valentine, J. W., and T. D. Walker. 1987. Extinctions in a model taxonomic hierarchy. *Paleobiology* 13:193–207.

Van Valen, L. M. 1985. How constant is extinction? *Evolutionary Theory* 7:93–106.

Vermeij, G. J. 1977. The Mesozoic marine revolution: Evidence from snails, predators, and grazers. *Paleobiology* 3:245–258.

Vermeij, G. J. 1983. Shell-breaking predation through time. In M. J. S. Tevesz and

P. L. McCall, eds., *Biotic Interactions in Recent and Fossil Benthic Communities,* pp. 649–669. New York: Plenum.

Vermeij, G. J. 1987. *Evolution and Escalation: An Ecological History of Life.* Princeton: Princeton University Press.

Walter, M. R. Du, R., and R. J. Horodyski. 1990. Coiled carbonaceous megafossils from the Middle Proterozoic of Jixian (Tianjin) and Montana. *American Journal of Science* 290-A:133–148.

7 : The Meaning of Systematics and the Biodiversity Crisis

Michael J. Novacek

The biological world is a delicate web of connections, a network of species reproducing, growing, feeding on one another, competing for the same resources, evolving, and ultimately going extinct. The building blocks of these finely tuned and, as we are increasingly aware, highly vulnerable systems, are the species themselves. Systematics is the science that is central to the problem of understanding the diversity of these species and the impact of their current and projected extinction on the well-being of the planet.

The forgoing paragraph is meant to underscore the importance of systematics. It would be difficult to imagine a more noble object of discovery than the organization of the living planet. Yet, the truths scientists may hold self-evident are sometimes esoteric and unsatisfying to a much broader audience. The nonpractitioners might well ask: what is systematics? Why is systematics important to the well-being of the planet? For that matter, why does a preoccupation with naming organisms and labeling specimens mean much of anything to the promotion of general knowledge or to the well-being of the planet?

To systematists engulfed by the excitement of their own mission of discovery, these questions can be distracting and disturbing. The questions not only mirror a detachment from the working of science but also reveal misconceptions about what systematists really are. Nevertheless, the questions, whether born of unfamiliarity or misconception, are perfectly justified. Sadly, there is no reason in our society why the true nature of systematics should be widely understood. In many high school and college textbooks, systematics is portrayed with some dispatch as that area of science dealing with the naming and classification of organisms. Little is said about what such classifications mean. Do they reflect the overall distinctiveness of the classified groups? Do they reflect strict patterns of descent from common ancestors? Do they reflect the degree to which organisms differ or the degree to which they are the same? Do they reflect the genetic fabric of organisms as well as their anatomy? Moreover, little is said about what actually happens when one does systematics. In fact, one is left to infer that the classifications of organisms are arrived at simply through the labeling of specimens.

If college textbooks and introductory courses often display these shortcomings and distortions, one can hardly expect more effective exposure at broader levels. Hence, the popular image of "the systematist as drawer-puller" endures. While it is true that many systematists may pull out many drawers and label myriad specimens during their lifetime, to be totally consumed by that practice would be a fatal course for any intellectual endeavor. If systematists were concerned only with labeling specimens, then there would be little chance to convey a sense of importance about systematics that extends beyond amassing collections or publishing great catalogues. Of course, systematics is much more than labeling specimens. The popular image simply overemphasizes that element of work referred to in the scientific parlance as "data collecting." The image of labelmaker obscures the powerful theoretical side to systematics that holds the biological sciences together.

Systematists must therefore be prepared to explain what they do in a language that can be appreciated by those who are not likely to share their vocation. A positive first step is to deal with the conundrum arising from the common usage of two words, *systematics* and *taxonomy*. These terms have been used variously and interchangeably. In my own experience, the word *taxonomy* to the uninitiated has the pedestrian connotation of a theoretically barren practice akin to cataloguing or naming things. The word *systematics,* on the other hand, often either means the same thing as taxonomy or means hardly anything at all. The dictionary is not much help here; *systematics* often receives the definition, "the science or a method of classification; esp., *same as* TAXONOMY" (*Webster's New World Dictionary,* Second College Edition, p. 1445). To those more familiar with biology, taxonomy and systematics may carry somewhat different meanings, but there is no dogma that establishes a universal definition for either word. The famous evolutionary biologist George Gaylord Simpson (1961:6–11) contributed some thought to the distinction between systematics and taxonomy. I here suggest definitions that, with minor retooling, reflect Simpson's:

Systematics is the science that deals with the organization, history, and evolution of life. It ultimately asks, how did life forms originate? How did they diversify and how are they distributed both in space and time?

Taxonomy is the practice of describing and naming life forms and arranging them in classifications that reflect patterns of relationships. It provides the language for systematics. It is a part of systematics, not the whole of systematics.

Taxonomy may thus be regarded as an operational core of systematics. Organisms must be named and described and their representative specimens catalogued so that a data bank of information is assembled. These taxonomic efforts require, however, a motive, and the motive is provided by the questions listed above under systematics. The questions are what bring a vast array of special interests together. Thus a paleontologist unearthing skeletons in an Asian desert and a molecular biologist sequencing a strand of deoxyribonucleic acid (DNA) can both claim to be systematists if they share an interest in how species are related and how they arose over time. All these issues depend on theories of patterns of descent, of organisms branching off from each other in a way that accurately reflects their histories. When such theories continue to successfully explain new observations, they form the basis for many statements about the biological world. Whether DNA evolves faster or slower in certain organisms is a systematic issue. Whether man has a stronger affinity with certain primates than with others is a systematic issue. Whether the slave-making societies of certain ant species are highly advanced or extremely primitive is a systematic issue. Whether most extinction is episodic or randomized throughout the history of life is a systematic issue. Whether life on earth arose once, twice, or many times is a systematic issue.

How does systematics address these kinds of questions? To answer, let us consider the current investigations of the earliest history of life. Suppose, for example, we wish to determine the conditions under which life evolved. Clues to solving this problem come from the study of bacteria, single-celled organisms that are known from rocks as much as three and a half billion years old. Accordingly, we might reconstruct the conditions for early life by examining the adaptations of living bacteria. At first glance, however, bacteria offer a complicated answer. Some bacteria are called methanogens because they derive their energy from the production of methane from carbon dioxide and hydrogen. Another group of bacteria are the halophiles, organisms that live in very salty environments because they require high salt concentrations for the production of energy. A third group are the thermophiles, bacteria that thrive only in water at or near boiling temperature and largely derive their energy from the metabolism of sulfur compounds. (Thermophilic bacteria are the microscopic organisms inhabiting the hot sulfurous pools at Yellowstone National Park.)

Which of these adaptations, if any of them, are most like those expected in the earliest life? We can approach this question through systematics only by reconstructing a family tree for the various kinds of bacteria. In constructing such trees based on differences in ribonucleic acid (RNA), systematists have found a very interesting pattern (Lake 1989; Woese 1989). Although methanogens and halophiles are restricted to a few branches of the tree, the thermophiles are widespread and appear in a number of major branches. Logic dictates that the adaptation that sorts out into the majority of these major branches can be most directly traced back to the central trunk of the tree. Although there is current debate on the precise branching pattern for bacteria (Lake 1989; Woese 1989) all sides seem to agree that the ancestors of primitive bacteria, more advanced bacteria, and even the higher organisms (protozoans, fungi, plants, and animals) were probably thermophilic. Indeed, there is agreement, spawned by our understanding of the systematics of these basic lineages, that all life probably arose in higher temperature environments. Such a conclusion has far-reaching implications. Not only does it offer a window into the earliest history of life on earth but it also provides an impetus for more detailed investigations of living thermophilic bacteria for chemical and cellular properties that might more nearly explain the emergence of life (Corliss 1990). Indeed, such a conclusion about the ancestry of life on earth could even influence our exploration for living organisms on other planets.

Given the universal nature of the issues, one would naturally expect that systematics forms its observations from a vast range of different kinds of evidence. It is again popular to portray systematists as consumed strictly with the details of body form through traditional techniques that involve such operations as counting the scales on a fish or the hairs on the back of an insect's leg. It is true that many decisions about the relationships of organisms depend on such basic information. But we also need to share with people the idea that the practice of *morphology,* or the study of body form and design, is a dynamic field armed with diverse and exciting new techniques. The electron microscope has opened new worlds of biological diversity. Experimental procedures have provided an inroad to understanding the relationship between form and function. Some current problems in morphology require sophisticated statistics, graphics, and even animation that challenge the world's most powerful computers.

Another dimension of systematic work not always appreciated is the range of applications outside the field of morphology. Although this morphological emphasis still dominates systematic work, it is important for systematists to stress there are many other ways in which evidence comes to light. These includes studies of development, physiology, behavior, and genetic change. One of the most exciting aspects of systematists today is the first opportunity to compare organisms for their genetic composition as well as their structure.

Moving from applications to objectives, we also find a disturbing lack of communication. Even biologists often have a poor sense of the theoretical issues that imbue systematics. Consider the central, first-level goal of systematics, namely, a theory for the history of relationships among a set of species or groups that contain species. These theories go by several names that mean much the same thing: *genealogies, phylogenies,* or *trees.* How one collects and analyzes such data is, in the textbook treatment, a straightforward matter, perhaps a matter of taste. But this could not be further from the truth. The methods for development of phylogenies represent a challenging and vigorously debated area of science. An ecologist colleague required all his graduate students to develop such phylogenies for the species that they were studying in the field. Most of these students could not believe how difficult this "preliminary" procedure in reality became. Several of the students were in fact so stimulated by this stage of the process that they never got to the ecological part of the study; they became full-fledged systematists. It is very safe to claim that systematics, especially during the last decade, represents one of the most theoretically sophisticated and intellectually vibrant areas of biology. Systematics, for example, represents the first bridge between the bold new discoveries in molecular biology and the more traditional fields emphasizing the biology of the whole organism.

How does the development of systematic theories, or phylogenies, relate to classification? Classifications should, after all, carry more meaning than a list of names for the sake of convenience. Above I listed a series of questions about classifications that are usually left unanswered in basic biology texts. The assumption is that if a classification "makes some sense," its genesis is of little concern. This ignores the fact that some classifications are counterintuitive—in some new classifications, for example, birds are simply a subgroup of dinosaurs. Yet it is hard to visualize the close kinship between a robin and *Tyrannosaurus rex.* Why then do such odd arrangements exist? The answer is that such classifications represent deep-seated patterns of relationships, or phylogenies, that are part of the process of discovery in systematics. In truth, classifications variously represent all items such as the degree of differences, the level of distinctiveness, or the pattern of kinship among organisms. In recent years there has been a commendable effort to bring the language of systematics represented by classifications more in line with the theories of systematics represented by phylogenies. This has not erased all the controversy, but science thrives on a cycle that alternates between controversy and enlightenment.

As for the importance of systematics in the agenda for the future of the planet, never has the message been so clear. The 1.4 million described species hardly qualify as an exhaustive dictionary of life. An educated guess concerning the actual number of species ranges between 5 and 80 million. It may seem odd that modern biology remains so ignorant. How many more parrot species are likely to turn up in your average acre of rain forest? How many more kinds of lions and tigers and bears lurk beneath the trees? The answer is, of course, that 5 to 80 million more species do not mean 5 to 80 million more of the furry and feathery creatures that have received most of our attention. The 5 to 80 million more species mean 5 to 80 million more tiny aquatic plants and animals, round worms, beetles, and various and sundry organisms valiantly studied by a comparatively thin rank of biologists—5 to 80 million more species whose doubtless importance to ecological communities is sadly unknown. Even more disturbing is another statistic—human activity may promote the extinction of nearly one fourth of the world's plant and animal species within fifty years. This is a very large number—namely, one fourth of 5 to 80 million species. The biological world is evaporating before we get a chance to effectively know it.

This decimation is understandably viewed as catastrophic by biologists. Its impact on our everyday lives and even on our health and welfare is also shockingly profound. Myriad species are beneficial to humankind in controlling pests and the spread of diseases. Human beings have the dubious distinction of promoting the extinction of key species and thus destroying the ecological relationships necessary for their own comfort and survival. The misfortune of some of these species may not raise the same outcry as the fate of the blue whale or the snow leapord. Yet the loss of any species has potentially detrimental effects. Few find bats particularly endearing, but certain species of bats, whose roosting habitats are currently being destroyed at an alarming rate, are the major large predator of night-flying insects (Constantine 1970). In regions like the southwest United States, where the formerly huge numbers of bats are diminishing, the prospects for control of agricultural pests are very bleak. Spiders are likewise a major land predator of many insects, including many actual and potential insect pests. Their importance in this regard is well evident in southern South America, where the diversity of spider species seems greater than even in tropical rain forests (Platnick, this volume). It is thus alarming that the very limited forested regions representing important spider habitats in this part of South America are being destroyed at a much greater proportional rate than the Amazon basin.

The litany of eliminated or endangered species need not be repeated here. If it is obvious that the loss of such organisms has grave bearing on the human condition, one might still ask how systematics offers insights and solutions relevant to human beings and the well-being of the planet. Aside from the noble mission to explain the history of life, systematics has a very practical and immediate application. The methods integral to systematics allow us to distinguish beneficial fungi (penicillium, edible mushrooms) from highly lethal and agriculturally detrimental members of this kingdom. For example, a notable number of deaths through mushroom poisoning are due to the consumption of species that superficially resemble highly edible forms. Discrimination of toxic and edible forms is based on detailed keys (see Krieger 1967) arrived at through careful systematic study. The courses of many tropical diseases have been traced by the host-parasite relationships in species that must first be systematically separated from closely related but harmless species. Insights on the source and identity of the human immunodeficiency virus (HIV) hinge on an understanding of the similarities and evolutionary relationships of the virus group to other viruses, namely, the very objectives that commonly drive systematic research. Not surprisingly, the *modus operandi* applied in this facet of HIV research is precisely the same as that in core systematic research (Hobson and Myers 1990). Recently, DNA evidence of Lyme disease was identified in a museum collection of ticks made several decades ago (Pershing et al. 1990). The systematists who originally collected, described, and catalogued these specimens were not on the track of Lyme disease; yet their expertise and efforts provide important evidence that the emergence of Lyme disease predated by at least a decade the formal recognition of this disease as a clinical problem in the United States. The application of systematics has practical benefits even in the case of extinct organisms; the discovery of great reservoirs of oil in marine rocks is often made from the taxonomic identification of tiny fossil organisms by systematic experts.

Systematics is also essential in our confrontation with the biodiversity crisis. We cannot hope, even with the best of resources and the largest army of biologists, to make a full accounting of this eroding biodiversity. Moreover, there must be scientific questions and objectives to constrain the process of discovery. Here, systematics plays a key role. It can identify priorities for the study of biodiversity. A recent recommendation, for example, places emphasis on the endangered species that are vital to our understanding of a broad picture of evolutionary history (May 1990). Another area where systematics can help in establishing priorities is in identifying unique species that are confined to very small and vulnerable corners of the earth. It is systematics that enables us to identify a unique Madagascar fauna or a bizarre flora from Baja California. It is systematics that will define the parameters of yet undiscovered biological communities in the tropical basins or the deep sea floor. A cornerstone to the study of the future of the world involves the description of its precious components. This involves the description and comparison of organisms for the development of phylogenies, evolutionary theories, and classifications—the area of science known as systematics. In order to consider how organisms originate, reproduce, grow, compete for resources within their communities, and ultimately go extinct, we must build on systematics, on an understanding of the organization of biological life. We must know about species and where they live before we learn of their biological roles and their importance to us and before we can seek effective ways of ensuring their survival.

Finally, we should also not expect all luminous answers from a science in its exploratory stages. There is an element here—beyond the theory and sophisticated technique—that simply say we should know what we are trying to explain. We should explore a world simply because "it is there." The age of exploration of biological diversity began when Greek philosophers provided a basic roll call of organisms and informed us, among other things, that whales were not fish. Dozens of centuries later, the age of exploration surged ahead with the efforts of John Ray and Carrol Linneaus and their classifications of a portion of the earth's species. We have since split the atom, landed on Mars, and deciphered the genetic code. Ironically, with only 1.4 million species recorded, many in only the most superficial terms, and perhaps as many as 80 million to go, the age of exploration of the biological world has barely begun.

ACKNOWLEDGMENTS

I thank Niles Eldredge for his invitation to contribute to the volume and for his editorial efforts. Vera Novacek and Mark Norell provided helpful comments on the manuscript.

REFERENCES

Constantine, D. G. 1970. Bats in relation to the health, welfare, and economy of man. In W. A. Wimsett, ed., *Biology of Bats*.

Corliss, J. B. 1990. Hot springs and the origin of life. *Nature* 347:624.

Hobson, S. W. and G. Myers. 1990. Human immunodeficiency viruses: Too close for comfort. *Nature* 347:18.

Krieger, L. C. C. 1967. *The Mushroom Handbook*. New York: Dover.

Lake, J. 1989. Origin of eukaryote nucleus determined by rate-invariant analyses of ribosomal RNA genes. In B. Fernholm, K. Bremer, H. Jornvall, eds., *The Hierarchy of Life*, pp. 87–103. Amsterdam: Elsevier.

May, R. M. 1990. Taxonomy as destiny. *Nature* 347:129–130.

Pershing, D. H., S. R. Telford III, P. N. Rys, D. E. Dodge, T. J. White, S. E. Malawista, A. Spielman. 1990. Detection of *Borrelia burgdorferi* DNA in museum specimens of *Ixodes dammini* ticks. *Science* 249:1420–1423.

Simpson, G. G. 1961. *Principles of Animal Taxonomy*. New York: Columbia University Press.

Woese, C. R. 1989. Archeobacteria and the nature of their evolution. In B. Fernholm, K. Bremer, H. Jornvall, eds., *The Hierarchy of Life*, pp. 119–130. Amsterdam: Elsevier.

8 : Phylogenetic Analysis and the Role of Systematics in the Biodiversity Crisis

Melanie L. J. Stiassny

The total array of the earth's biological diversity will probably never be fully known. Species and whole communities are being lost even before they can be identified, and the richness of the planet is irreversibly diminished. It is in the face of this profound crisis of biodiversity and the complex economic, social, and ethical problems it engenders that biologists are faced with the onerous task of allocating conservation priorities. If there were some absolute biological criterion with which to judge the relative value or worth of a given species or community against another, perhaps the problem would not be so acute, but unfortunately no such criterion exists. So how are conservation priorities being established and does systematics play a particular role in this process?

Even a brief review of the current conservation literature reveals that many different criteria are being used in assessing the conservation potential and/or ecological value of communities, areas and even individual species. In total, these reflect a broad sweep of conservation priorities ranging from the preservation of rare species or fragile environments to the maintenance of diversity and community stability, and the protection of representative samples of regional and global ecosystems and biotas. Indices such as representativeness, diversity (including both species richness and habitat diversity), species-area relationship, naturalness (in the sense of minimal human disturbance), and rarity are widespread and commonly used. Underpinning each of them is the clear need for a comprehensive knowledge of the taxonomic composition and the biogeographical and historical distributions of the various entities under consideration. Without this basic information, no inventories of threatened areas, endangered species lists, extinction projections, or programs for rational planning are possible. In this sense, systematics clearly plays a key role in conservation biology. Without this fundamental systematic data we would be unable to document the fact that we are in the midst of a crisis of biodiversity. Taxonomic and biogeographical analysis, two major components of systematic biology, are clearly fundamental elements of conservation biology also.

There is another aspect to systematic biology that has yet to be fully incorporated into conservation biology. This missing element is phylogenetic analysis, and the question of what role there may be for phylogenetics in conservation biology has only recently begun to be addressed (e.g., Vane-Wright et al. as cited in May 1990; May 1990). It is this question of the potential role of phylogenetic analysis, particularly as it relates to informing our choices in designating conservation priorities, that I would like to explore further here.

Phylogenetic Analysis

Essentially concerned with the reconstruction of evolutionary history, phylogenetic analysis seeks to determine the genealogical interrelationships of taxa. By using the distribution of intrinsic properties of organisms (characters), lineages and their composition and interrelationships are determined and depicted in the form of branching diagrams. Phylogenetic trees portray the genealogical relationships and sequence of historical events uniting taxa and form the baseline for virtually all comparative evolutionary studies. Do they have a role to play in conservation biology?

To examine this question further, I shall emphasize as an example studies of the Cichlidae, a family of freshwater fish with which I am familiar.

The Cichlidae are one of the most speciose and ecologically diverse families of spiny-rayed fishes. They are primarily fishes of the lowland tropics with a widespread natural distribution conforming to an essentially Gondwanan pattern (fig. 8.1). Currently, the oldest (unambiguously) cichlid fossils appear in the Oligocene of Africa and South America (van Couvering 1982). These early cichlids are clearly morphologically derived and it is clear that the origin of the family long predates its earliest fossil record. Much interest has centered on the evolutionary dynamics of this speciose family. Most recently the spectacular demise of the cichlid radiations of Lake Victoria, East Africa (Barel et al. 1985; Ogutu-Ohwayo 1990), has helped focus the attention of conservationists on the plight of these fragile aquatic communities.

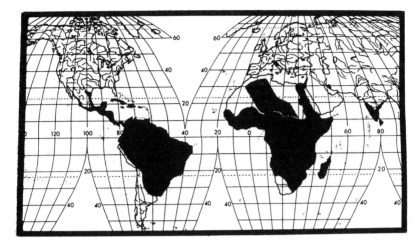

Figure 8.1 Distribution map for the family Cichlidae (modified after Berra 1981). Light gray patch across the Sahara Desert indicates scattered records in isolated waterholes, etc.

An overview of the phylogenetic intrarelationships of the major cichlid lineages is presented in figure 8.2A. When the broad geographical distribution of clades and the approximate numbers of species included within them are superimposed onto this tree (fig. 8.2B), a number of interesting points arise. Even at the coarse scale depicted here, it is immediately obvious that cichlid species are not uniformly arrayed over their familial range. For example, Africa has a vast number of species when compared with the cichlid assemblages of the other continents. Conversely, Madagascar, with its large landmass and complex hydrography (Aldegheri 1972), has a notably depauperate cichlid fauna with only nine extant species known. Certainly if species richness were to be the criterion of choice, the Madagascan Cichlidae could almost be described as insignificant. After all, in central east Africa, Lake Malawi alone may harbor anywhere between 500 to 1000 species of cichlid fish (Lewis, Reinthal, and Trendall et al. 1986). At first sight, surely most would agree that the answer to the grim question, which shall we conserve, the Madagascan 9 or the Malawian 1000? would be to save the latter.

Phylogenetic Criteria

Norman Platnick (this volume) has addressed the question of what could be termed "maximizing species numbers." He correctly points to the critical issue of species endemism and the important biohistorical information con-

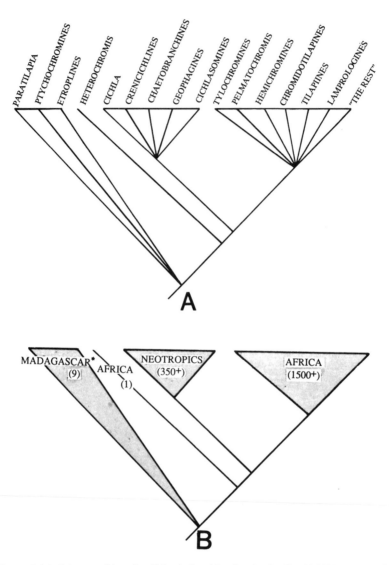

Figure 8.2A Scheme of intrafamilal relationships for the family Cichlidae. For a list of genera included in each of the named lineages, and characters defining the various assemblages, see Stiassny 1990b, 1991. B. Major cichlid clades with broad geographical distributions and approximate species numbers superimposed onto the scheme of relationships. *Three additional etropline species are endemic to southern India and Sri Lanka.

tained in narrowly distributed taxa. However, in my example, degrees of endemism are not the issue, for both the Madagascan and Malawian radiations are narrowly endemic. Here we have more of a "quality versus quantity" dilemma, and it is in cases such as this that other historical factors may inform our decisions. It is these additional factors that I am terming the "phylogenetic criteria."

Figure 8.2A illustrates that although the precise relationships of the Madagascan assemblage are yet to be fully resolved (see also Stiassny 1991), they represent the phylogenetically most plesiomorphic cichlid lineage(s) and, as such, occupy a key position with respect to the rest of the family. Of course, in representing the oldest of cichlid lineages, they certainly have had a long time to accrue a range of (autapomorphic) specializations and in that sense have probably diverged markedly from the actual ancestors of the family. However, perhaps more importantly, they lack precisely that series of specializations (morphological, biochemical, behavioral, or otherwise) that characterize the more derived cichlid clades. Why is this important?

Much has been written about the ecological and evolutionary questions posed by the existence of species-rich, adaptively multiradiate, and narrowly endemic communities of cichlids, and these fish have come to occupy an important place in modern evolutionary studies (e.g., Fryer and Iles 1972; White 1978; Futuyma 1979; Stanley 1979; Vrba 1980; Greenwood 1984). To evaluate hypotheses concerning the operation of evolutionary mechanisms and processes such as modes and rates of speciation and the acquisition and role of evolutionary novelties, a corroborated theory of phylogenetic relationships is of critical importance (Eldredge and Cracraft 1980; Nelson and Platnick 1981; Wiley 1981; Lauder 1982). For the Cichlidae, at least at the level I am considering here, such a broad theory is available (fig. 8.2A). While we know nothing of the direct ancestry of the family (Stiassny and Jensen 1987), as figure 8.2A indicates, in the Madagascan taxa we have, still living, representatives of the most basal of cichlid lineages. These phylogenetically basal Madagascan taxa should be of particular interest to evolutionists for they provide a unique resource for baseline comparative studies within the family. This is true, not only for our understanding of the temporal sequence of acquisition and the possible role of morphological innovation in taxic diversification (for which data from fossil material may one day become available), but also for behavioral studies. For example, the potential role of complex mating systems and sexual selection in the speciation processes of cichlid fishes has repeatedly been invoked (see review by Dominey 1984). The phylogenetic position of the Madagascan clades renders an analysis of their reproductive biology of particular interest. As long as the Madagascan taxa remain extant, the possibility to investigate it, and perhaps thereby gain valuable insight into the evolution of sexual systems within the family, remains an option (Stiassny and Gerstner, in press).

By the foregoing I do not mean to imply that the truly spectacular East African lacustrine radiations of cichlid species are of only minor interest or importance in evolutionary studies or that little effort should be expended on their conservation. I want instead to make the rather simple point that from a phylogeneticist's viewpoint the loss of nine Malawian species, for example, is not equivalent to the loss of the nine Madagascan species. When viewed from this perspective not all species are created equal.

Phylogenetic analysis reveals other interesting parallels. For example, the endemic Madagascan silversides of the family Bedotiidae, another rather depauperate clade of only six species, exhibit an identical pattern of relationship with respect to the remaining atherinomorph radiation (see figure 8.3). Similar claims may also be made for a phylogenetically basal placement of the endemic freshwater Madagascan mullet of the genus *Agonostomus* (see Stiassny 1990a), and possibly also the endemic freshwater catfishes of the genus *Ancharius* (Ferraris, in preparation). Interestingly, as with the majority of the Madagascan Cichlidae, these other taxa are restricted in their distributions to the eastern forests of the island.

At least for a number of the freshwater ichthyofaunal components, it ap-

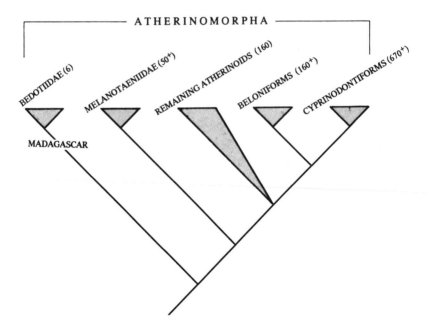

Figure 8.3 Atherinomorph intrarelationships. For details of characters defining assemblages, see Stiassny 1990a.

pears that these phylogenetic data may be used to bolster the argument for a broad regional conservation plan, in this case recognizing eastern Madagascar as an area harboring taxa of considerable phylogenetic interest. In this particular example the proposal is more or less equivalent to highlighting areas of endemism as candidates for conservation priority, but this need not always be the case.

Returning to the phylogenetic tree for the family Cichlidae, another taxon of obvious interest is the genus *Heterochromis,* represented by the single species *Heterochromis multidens* (fig. 8.2B). This poorly known taxon was first described in 1900 by Pellegrin. Its relationships were obscure, but most subsequent authors considered *Heterochromis multidens* to be a tilapiine of some sort, and as such, just another of the numerous African cichlid taxa. Oliver (1984) was the first to recognize the primitive nature of this fish and suggest that it represented the most phylogenetically basal of all African cichlids. My own studies of the species (Stiassny 1987, 1991) indicate that *Heterochromis multidens* is in fact the sister taxon to the remaining Afro-Neotropical radiation (fig. 8.2A). This is a somewhat unexpected finding and raises some interesting questions with respect to the age of the family and the pertinence of Gondwanan fragmentation models in explanations of intrafamilial diversification.

Heterochromis multidens has a restricted distribution, being found only in certain forested backwater habitats of the *cuvette centrale* of the Zaire River (fig. 8.4). Virtually nothing is known of the basic biology or ecology of the species and it is poorly represented in museum collections. The limited distribution and close association of *Heterochromis multidens* with rainforested biotopes suggest that, like so many other rainforest-adapted species, this fish is probably extremely vulnerable to deforestation pressure (Reinthal and Stiassny 1991).

In the event of the not unlikely scenario that *Heterochromis multidens* were to be threatened with extinction, a strong argument for the implementation of some form of conservation measures could obviously be made on the grounds that this is an endemic species from a region of high endemicity (the Zaire basin harbors 690 fish species of which 80% appear to be endemic [Lowe-McConnell 1987]). However, in the case of *Heterochromis multidens,* the argument for conservation priority is strengthened by knowing the phylogenetic position of the taxon. The loss of this species is not commensurate with the loss of "just another" of the numerous African cichlids, and it is phylogenetic analysis that indicates why this is the case.

In the foregoing discussion I have indicated that phylogenetic analyses, like the other components of systematic biology, have an important role to play in conservation biology. Some taxa, by virtue of their basal phylogenetic position, are of particular importance in our understanding and interpretation

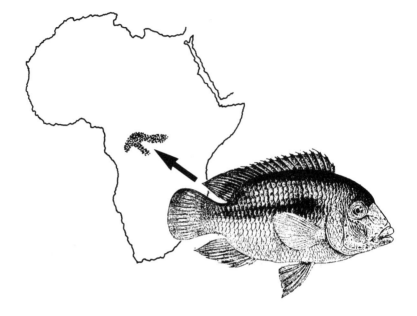

HETEROCHROMIS

Figure 8.4 Geographical distribution of *Heterochromis multidens* (Pellegrin 1900). Illustration of *Heterochromis multidens* after Boulenger 1915.

of the evolutionary history of the groups to which they belong. Frequently these basal lineages are also taxonomically depauperate and have extremely limited geographical distributions. The loss of these taxa has a particular resonance, often resulting in the extinction of entire evolutionary lineages.

Phylogenetic arguments such as those outlined here are essentially qualitative, but the beginnings of a more quantitative application of phylogenetic criteria are being developed. For example Vane-Wright et al. (cited in May 1990) have proposed a system that seeks to extract an index of "taxonomic distinctness" from phylogenetic trees. This procedure is outlined by May (1990) and is illustrated here by reference to the cichlid family tree depicted in figure 8.5A. If one counts nodes from each assemblage of species (A, B, C, and D) to the base of the tree, four counts are obtained: 1 node for A, 2 nodes for B, and 4 nodes each for C and D. Each assemblage may now be assigned an index of taxonomic distinctness that is inversely proportional to that node count. By scaling the minimum index to 1, the values for species in each of the assemblages becomes: 4/1 = 4 for A, 4/2 = 2 for B, and 4/4 = 1 for C and D respectively.

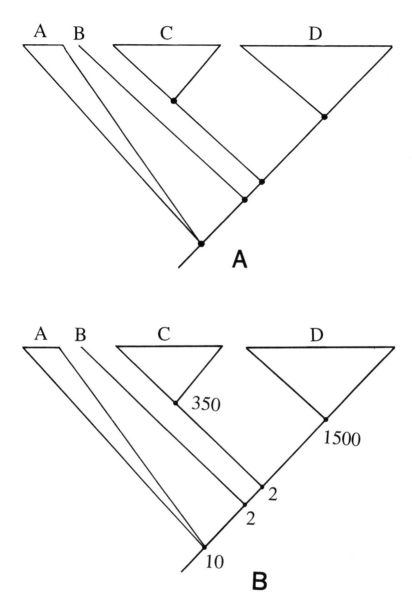

Figure 8.5A Topography of cichlid family tree. B. Cichlid family tree with approximate numbers of branches at each node indicated.

May (1990) has proposed a modification of this system that incorporates a consideration of how many lines branch from each node on the tree. Following May's (1990) procedure, again with reference to the cichlid tree (figure 8.5B) a modified index of taxonomic distinctness for any species in each of the cichlid assemblages (A, B, C, or D) can be calculated by totaling branch numbers at summed nodes and scaling to 1 as follows: $1514/10 = 151.4$ for A, $1514/12 = 126.17$ for B, $1514/364 = 4.16$ for C, and $1514/1514 = 1$ for D. In this example, the order of distinctiveness of the different assemblages remains unchanged, but May's modified index strongly emphasizes the extreme distinctness of each of the basal and taxonomically depauperate lineages highlighted by this study, i.e., an index of 151.4 for the Madagascan lineages, and 126.17 for *Heterochromis multidens*.

In an ideal world we would, of course, argue to save everything, but the world we live in is far from ideal. The rachet of extinction clicks faster each day and our resources seem to diminish in inverse proportion. Systematists, like everyone else, have an agenda to fight for. If we are going to succeed we need to express ourselves in a way that is both accessible and convincing to specialists and nonspecialists alike. Happily, there is a growing recognition of the vital role that museums and museum-based biological research must play in monitoring the biodiversity crisis (e.g., Wilson 1985, 1988; Barrowclough, this volume). There is an obvious need to know what is out there, how much there is of it, and how endangered it is, and as systematists we should have little problem in convincing the conservation community and the "public at large" of the central importance of our science. The potential role of phylogenetic analysis in helping to define conservation priorities is perhaps less immediately obvious but is also of considerable importance.

REFERENCES

Aldegheri, M. 1972. Rivers and streams on Madagascar. In J. Illies, ed., *Monographiae Biologicae* 21:261–310. The Hague: Dr. W. Junk.

Barel, C. D. N., R. Dorit, P. H. Greenwood, G. Fryer et al. 1985. Destruction of fisheries in Africa's lakes. *Nature* 315:19–20.

Barrowclough, G. F. In this volume.

Berra, T. M. 1981. *An Atlas of the Distribution of Freshwater Fish Families of the World.* Lincoln: University of Nebraska Press.

Boulenger, G. A. 1915. *Catalogue of the Fresh-Water Fishes of Africa in the British Museum* (Natural History), vol. 3. London: Trustees of the British Museum (Natural History).

Dominey, W. J. 1984. Effects of sexual selection and life history on speciation: Species flocks in African cichlids and Hawaiian *Drosophila*. In A. A. Echelle and I. Kornfield, eds., *Evolution of Fish Species Flocks*, pp. 231–249. Orono: University of Maine at Orono Press.

Eldredge, N. and J. Cracraft. 1980. *Phylogenetic Patterns and the Evolutionary Process: Methods and Theory in Comparative Biology.* New York: Columbia University Press.

Fryer, G. and T. D. Iles. 1972. *The Cichlid Fishes of the Great Lakes of Africa. Their Biology and Evolution.* Edinburgh: Oliver and Boyd.

Futuyma, D. J. 1979. *Evolutionary Biology.* Sunderland, Mass.: Sinauer Associates.

Greenwood, P. H. 1984. African cichlids and evolutionary theories. In A. A. Echelle and I. Kornfield, eds., *Evolution of Fish Species Flocks,* pp. 141–154. Orono: University of Maine at Orono Press.

Lauder, G. V. 1982. Historical biology and the problem of design. *Journal of Theoretical Biology* 97:57–67.

Lewis, D., P. Reinthal, and J. Trendall. 1986. *A Guide to the Fishes of the Lake Malawi National Park.* Gland, Switzerland: world wildlife fund.

Lowe-McConnell, R. H. 1987. *Ecological Studies in Tropical Fish Communities.* Cambridge, U.K.: Cambridge University Press.

May, R. M. 1990. Taxonomy as destiny. *Nature* 347:129–130.

Nelson, G. and N. Platnick. 1981. *Systematics and Biogeography: Cladistics and Vicariance.* New York: Columbia University Press.

Ogutu-Ohwayo, R. 1990. The decline of the native fishes of Lakes Victoria and Kyoga (East Africa) and the impact of introduced species, especially the Nile perch, *Lates niloticus* and the Nile tilapia, *Oreochromis niloticus. Environmental Biology of Fishes* 27:81–96.

Oliver, M. K. 1984. Systematics of African cichlid fishes: Studies on the haplochromines of Lake Malawi (Teleostei: Cichlidae). Ph.D. dissertation, Yale University, New Haven, Conn.

Pellegrin, J. 1900. Poissons nouveaux du Congo francais. Bulletin de Muséum Nationale d'Histoire Naturelle, Paris 6:98–101.

Platnick, N. In this volume.

Reinthal, P. N. and M. L. J. Stiassny. 1991. The freshwater fishes of Madagascar: A study of an endangered fauna with recommendations for a conservation strategy. *Conservation Biology* 5(2):231–243.

Stanley, S. M. 1979. *Macroevolution. Pattern and Process.* San Francisco: W. H. Freeman.

Stiassny, M. L. J. 1987. Cichlid familial intrarelationships and the placement of the neotropical genus *Cichla* (Perciformes, Labroidei). *Journal of Natural History* 21:1311–1331.

Stiassny, M. L. J. 1990a. Notes on the anatomy and relationships of the bedotiid fishes of Madagascar with a taxonomic revision of the genus *Rheocles* (Atherinomorpha: Bedotiidae). *American Museum Novitates* 2979:1–33.

Stiassny, M. L. J. 1990b. *Tylochromis,* relationships and the phylogenetic status of the African Cichlidae. *American Museum Novitates* 2993:1–14.

Stiassny, M. L. J. 1991. Phylogenetic intrarelationships of the family Cichlidae: an overview. In M. H. A. Keenleyside, ed., *Cichlid Fishes: Behaviour, Ecology, and Evolution.* London: Chapman Hall.

Stiassny, M. L. J. and J. S. Jensen. 1987. Labroid intrarelationships revisited: Morphological complexity, key innovations, and the study of comparative diversity. *Bulletin of the Museum of Comparative Zoology* 151(5):269–319.

Stiassny, M. L. J. and C. L. Gerstner. In press. The parental care behaviour of *Paratilapia polleni* Bleeker, 1868 (Perciformes, Labroidei) a phyogenetically primitive cichlid from Madagascar, with a discussion of the evolution of maternal care within the family *Cichlidae. Environmental Biology of Fishes.*

Van Couvering, J. H. A. 1982. Fossil cichlid fish of Africa. *Special Paper of the Palaeontological Association* 29:1–103.

Vrba, E. S. 1980. Evolution, species and fossils: How does life evolve? *South African Journal of Science* 76:61–84.

White, M. J. D. 1978. *Modes of Speciation*. San Francisco: W. F. Freeman.

Wiley, E. O. 1981. *Phylogenetics. The Theory and Practice of Phylogenetic Systematics*. New York: Wiley.

Wilson, E. O. 1985. The biological diversity crisis: A challenge to science. *Issues in Science and Technology* 2:20–29.

Wilson, E. O. 1988. The current state of biological diversity. In E. O. Wilson, ed., *Biodiversity,* pp. 3–18. Washington, D.C.: National Academy Press.

9 : Systematics, Biodiversity, and Conservation Biology

George F. Barrowclough

There are many aspects to the biodiversity crisis (Wilson 1988a; Freedman 1989; Silver 1990), and museum personnel have been involved, to varying extent, in most of them. It is not possible, however, to review the entire field and all the contributions, actual and potential, in this forum. Rather, this paper is primarily an attempt to examine some of the major conceptual issues in the crisis and to comment on how museum (and herbarium) professionals have contributed to solving these theoretical and empirical problems. In addition, I point out issues to which museum workers ought to contribute.

At the onset it is necessary to make some distinctions and try to impose some order so that a heterogeneous topic can be pursued expeditiously. One primary distinction is between the documentation and the conservation of taxa and diversity; a second major categorization concerns geographic scales. These structures have not been well defined in the popular press (e.g., Easterbrook 1990).

Geographic Scales

Concern about biodiversity covers three complementary geographic scales. First, there exist global problems that threaten entire environments of the earth (Lean, Hinrichsen, and Markham 1990). For example, the depletion of ozone in the upper atmosphere due to the accumulation of carbon-halogen compounds can occur globally although at present it has mainly been observed at high latitudes (Silver 1990). Global warming (Silver 1990) is a similar, widespread phenomenon, as is acid rain (Freedman 1989). Problems on this scale are largely due to human activities; for the most part they must be addressed internationally through political and social decisions. To some extent systematists may provide a service in identifying such large-scale forces: for example, research based on museum collections of eggshells of birds was instrumental in the discovery that some pesticides were responsible for the breeding failure of a number of disparate birds whose common feature was their position high in the food chain (Risebrough 1986). However, the role of the systematist is not primary at this global scale. To some extent it does seem fatuous to develop plans and create refuges to preserve various ecosystems in the face of systemic pressures irrevocably altering temperatures, rainfall patterns, and penetration of the atmosphere by ultraviolet radiation. Paleontologists and evolutionary biologists can point out how these problems may result in global extinction and can provide a historical perspective on these problems (e.g., this volume).

A second geographic scale of concern is a regional or biomic one. This is the scale at which most scientific research is necessary and the one at which systematists can make the greatest contribution; it is the major subject of this review and of the discipline known as conservation biology. Before pursuing this aspect in detail, however, it is necessary to comment on a third geographic scale of interest, the local one.

Local biodiversity problems concern the preservation of species and communities in the local area, for example, in cities, counties, or states. For the most part such problems do not involve the loss of species; rather, the issue is the maintenance of the local environment for aesthetic and recreational interests. Consequently, a major aspect of the crisis at this scale is conflict between these aesthetic needs and development and other human activities. Because it is often difficult to place a satisfactory monetary value on parks and reserves, local conservation efforts often suffer (Norton 1987; Easterbrook 1990). Museum personnel, through their private activities, lecture series, and courses, and by maintaining field stations, have helped to raise the awareness of the general population concerning this issue. As an example, museum membership, lectures, and local birdwatching trips help create a concerned community supporting local environmental activism.

Conservation Biology

Museum researchers can make diverse contributions to the problems associated with the biodiversity crisis at several geographic scales; nevertheless, it does seem to me that their major impact as professionals ought to be in their area of expertise—systematics and allied concepts. In particular, the immediate important issue is the documentation and maintenance of native diversity on regional scales (Soulé 1990). This might be called "conservation biology," as opposed to "biodiversity politics" or "local community conservation action." The major issues in conservation biology are: (1) providing justification for documentation and preservation activities, (2) documenting diversity, (3) developing objectives and techniques for conservation management, (4) providing guidelines for the geographic placement of reserves, and (5) developing guidelines for eventual, difficult decision making in regard to alternate taxa and areas.

The Value of Biodiversity

There is already a large literature that discusses the logic and rationale for preserving biodiversity. For the most part, these arguments can be divided into economic advantages and ethical/aesthetic reasons (Norton 1987). Among the economic pluses are the possibilities of direct agricultural harvesting of native fruits, nuts, and other products, for example, latex, from relatively undisturbed habitats (Pinedo-Vasquez et al. 1990). Indirect agricultural value may result from the discovery of useful native plants or animals for eventual intensive farming whose usefulness has not yet been realized (Plotkin 1988); related is the potential for isolating genes for increased productivity, disease resistance, etc., from native plants (Iltis 1988; Plotkin 1988). Another possibility is the discovery of compounds of pharmaceutical value in plants and fungi (Lovejoy 1986; Farnsworth 1988). Finally, maintaining biodiversity may have immediate, direct economic consequences through ecotourism and other recreational activities; the game parks of East Africa and the national parks of the United States illustrate this possibility. However, both Norton (1987) and Ehrenfeld (1988) have pointed out that all such economic arguments are potentially treacherous: today's present value of even very large economic payoffs at an indefinite time in the future is apt to be quite low and hence not competitive with alternatives with immediate return.

The major alternative to economic arguments is the belief that species and diversity have intrinsic or transformative values that ought to be recognized per se (e.g., Callicott 1986; Norton 1987). This line of reasoning leads to value judgments and aesthetics that may not be universally accepted (e.g., the intrinsic worth of nonhuman objects, the value of the wilderness experience, etc.), particularly in developing societies (e.g., Radulovich 1990). However, systematists find an additional noneconomic value in studying diversity.

The major role for systematists in developing a claim for the value of studying, documenting, and preserving biodiversity is in arguing for the intrinsic value of knowledge and the value of a historical perspective. As Wheeler (1989) has pointed out in a related context, systematists need to make the argument that the primary product of systematic biology is a conceptual framework of evolutionary history. Conservation biology, in addition to economic and aesthetic benefits, will preserve the raw material needed for generating this vision and permit the development of a conceptual understanding and historic view of ourselves and our world. Systematists have not done a good job of articulating this line of reasoning. In a recent book devoted to the biodiversity crisis and edited by an entomologist (Wilson 1988a), for example, no one discussed the systematist's unique view. The value of the Amazonian forest is real, but amid its cachet, it is not clear that the public has any sense of how the floras and faunas of southeast Brazil, or even of Madagascar or New Zealand, inform us about our past and that of the planet.

The Documentation of Biodiversity: Patterns Within and Among Taxa

A prerequisite to making any decisions concerning the preservation of populations, species, or higher taxa is knowledge of their existence. Creating such knowledge is a dominant part of the research of systematists, and few other workers are sufficiently skilled to undertake such studies. This aspect of the current biodiversity crisis gives museum and herbarium workers an essential role that may not be as widely appreciated as it ought to be: there is a tendency for nonspecialists to think in terms of birds and larger mammals that are better known than most of the biota are. However, I believe even this is largely a misconception; for the most part species-level taxa are *poorly* known in most groups, including birds and mammals, and the studies of intraspecific variation as well as higher level relationships are all in their infancy.

Species

As others have observed, most of the species that presumably exist in the world are as yet undescribed (Wilson 1988a). If one does not know where they occur, how to identify them, or what their requirements are, then they can neither be documented for future study nor managed successfully and conserved as viable populations. As Platnick (this volume) points out, systematists are not even certain where the greatest biodiversity exists; it may not be in the tropics if spiders are representative of invertebrates. Thus, the essential role of systematists in studies of biodiversity and conservation biology is the cataloguing of diversity—formal description, identification, and documentation, and interpretation in the form of generalizations about relationships, biogeography, and endemism.

The inadequacy of current systematic knowledge may not be widely appreciated; nevertheless, it is real. There are two reasons that so much of our biota is not known: first, much of it has not yet been collected, studied, and formally described and named; second, there may be multiple, morphologically very similar appearing species considered single species because of insufficient study. The former is certainly the case for many invertebrates, especially arthropods. Erwin's (1988) widely known work on Neotropical beetles indicates the infancy of the discipline for insects; a minority of the species that he collected in forest canopy sites in lowland Peru had previously been described. On the basis of these and additional data, Erwin concluded that the world might contain as many as 30 million species of insects alone; fewer than one million have been described to date.

Knowledge of species and their numbers is also not complete even for better known taxa such as birds, mammals, fish, reptiles, and amphibians. This is because of inadequate sampling in part (e.g., Oren and Guerreiro de Albuquerque 1991) but also because of the existence of phenotypically identical (sibling) species. Until the recent past, systematists had to rely on morphology to detect specific differences. However, recent molecular studies have uncovered cases where morphologically identical or nearly identical populations have substantial molecular differences that suggest the existence of multiple taxa with long, separate evolutionary histories. This may be especially true in the tropics. For example, using protein electrophoresis, Capparella (1988) found species-level differences in a number of widespread, supposedly conspecific, Neotropical birds in terra firme forests on opposite sides of major rivers in Peru. Escalante-Pliego (1992) obtained a similar result in widely distributed marsh-dwelling birds of the genus *Geothlypis,* again in South America; her results indicate the existence of four allopatric species where one widespread taxon had been recognized. Even in temperate North America, however, newer molecular methods have led to the recognition of two or more species of birds where one had previously been thought to exist (e.g., Barrowclough and Gutiérrez 1990; Zink and Dittmann 1991). Revelations of this nature are not unique to birds, however. For example, in mammals (Myers and Patton 1989), frogs (Hillis 1988), and salamanders (Highton, Maha, and Maxson 1989), sibling species are still being identified at temperate and tropical latitudes. Thus, I conclude that, generally, the species-level taxa of the world are inadequately to very poorly known. For varying reasons, this is true of all taxa, mammals and birds as well as insects and spiders.

Intraspecific Variation

The study of geographic variation and intraspecific taxonomy is even more rudimentary than species-level taxonomy is. Only for a very few groups, such as some birds, mammals, commercially important fishes, and a few other taxa, have any extensive studies been performed. Although perhaps not as crucial as documenting unknown species, this work is nevertheless of importance in conservation biology and biodiversity issues for several reasons. First, as discussed above, there is the distinct possibility that some single, supposedly undifferentiated species actually include unrecognized sibling species. Second, species definitions are not uniform from taxonomic group to taxonomic group nor from worker to worker within groups. Thus, even in relatively well-known taxa, such as birds, the units of importance for conservation purposes may correspond to subspecies or races, and this may not be apparent without detailed studies of series of specimens from many localities (Barrowclough and Flesness 1992). For example, the widespread application of the biological species concept, with its emphasis on reproductive isolating mechanisms, has often resulted in the recognition of single species comprising multiple taxa, each of which has had its own separate evolutionary history. Some of these biological species may not even be monophyletic (e.g., Zink and Dittmann 1991). Third, although many studies of geographic variation will probably not reveal unrecognized species, they may uncover patterns in nature worth documenting and conserving. For example, clines are geographic gradients in characteristics of natural populations that reflect evolutionary processes, such as selection, gene flow, secondary contact, and recent range extensions, that are still not adequately understood (Endler 1977) but nevertheless represent a real and widespread aspect of the diversity of nature.

Higher Taxa

The inference of hierarchical patterns is one of the major research activities in museums; without extensive collections and knowledge of methodology, few others are able to undertake this task. Further, knowledge of hierarchical patterns is essential to any understanding of adaptation and the patterns and processes of evolution (e.g., Felsenstein 1985; Coddington 1988; Funk and Brooks 1990) and, as mentioned previously, for providing a historical perspective on biodiversity and the natural world (Eldredge and Cracraft 1980). But, in fact, many nonsystematists may not recognize the existence or value of higher taxa. Only recently have systematists started to write for a broader audience, scientific and general, to convey the idea that a deep understanding of biodiversity is dependent on knowledge of hierarchical, evolutionary relationships. In the biodiversity crisis in particular, knowledge of higher level relationships may affect important decisions about priorities in conservation effort (see below).

Collecting and Training

The major task of systematists in documenting the diversity of life is immense and barely begun. However, this role is not limited to offices, laboratories, and collections in museums and herbaria. In addition to the description and identification of species, specimens must be found and collected. This is also a task for specialists. Knowledge of how, when, and where to collect specimens requires considerable knowledge of the taxa of interest. A casual observer of birds or butterflies will not be a proficient collector of snakes, insects, and bryozoans. There are major roles for museum workers in field collecting and in training others to become adept at such work. The present distribution of systematists and taxonomists does not accurately reflect the abundance of organisms (figure 9.1); some taxa, such as vertebrates, are being studied by a disproportionate share of systematists while other taxa, such as arthropods, are relatively neglected. Extensive training of students, especially those of the developing world, is necessary to correct this imbalance. Other-

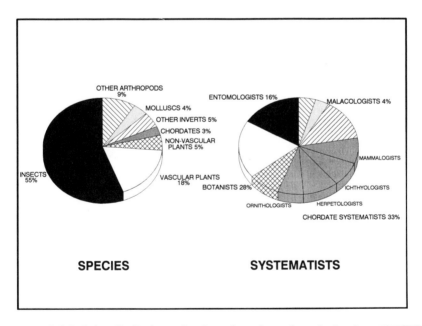

Figure 9.1 Relative distributions of estimated numbers of species-level taxa in the world and of the taxonomists who specialize on them. Distribution of taxa after Wilson (1988b); distribution of taxonomists based on numbers of systematists in the United States following Stuessy and Thomson (1981).

wise, patterns of distribution of birds and a few conspicuous mammals, butterflies, and plants will form the basis for generalizations about higher taxa.

The Mechanics of Preservation

Once it has been decided that some population, taxon, or habitat exists that is worth intervention and conservation, there is the problem of the details of preservation. That is, how does one design a program that will lead to the indefinite maintenance of diversity? This is a major field of research that is currently active, but the results to date are quite preliminary. Museum workers have been involved, but not dominant, in this effort. This is in part because the details of preservation often involve the design of refuges, techniques of wildlife management, and theoretical results in population dynamics and population genetics; many of these disciplines are peripheral to systematics.

Two alternate philosophies have arisen for dealing with the problem of preservation. In analogy to economics, I shall refer to these as macroconservation and microconservation theory. The former seeks to preserve entire communities through predictions based on empirical generalizations about biogeography while ignoring hard-to-measure details of population biology. The latter seeks management techniques for the preservation of an individual population or set of populations through the development of detailed models of its biology.

Macroconservation Biology

The macroconservation approach to the maintenance of biodiversity arose from empirical observations and the resulting generalization called the theory of island biogeography. This has been an active field since the appearance of a seminal monograph by MacArthur and Wilson (1967).

The theory of island biogeography is phenomenological. It results in attempts to predict patterns based on generalizations from empirical results that largely treat species as interchangeable elements. That is, large islands have more species of any given taxon than smaller islands do; hence, larger refuges will preserve more species than smaller refuges will (Sullivan and Shaffer 1975). Other common observations include the fact that islands close to a continental source have more species than isolated islands do. A scenario based on ideas of dispersal ability, population size, and extinction probability was elaborated in the form of graphic models that result in predictions in general agreement with observations on widely varying taxa (e.g., figure 9.2). This has led to a series of generalizations about how to design reserves for

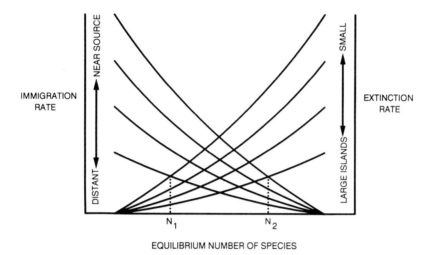

EQUILIBRIUM NUMBER OF SPECIES

Figure 9.2 Schematic representation of the equilibrium theory of island biogeography. Equilibrium number of species present on an island is determined by the intersection of immigration and extinction rates. The immigration rate of new species decreases as the number of taxa already present increases; the immigration rate is greater for islands close to a source of immigrants and lesser for islands distant from a source. The extinction rate increases with increasing number of resident species; the extinction rate is greater for small islands and lesser for large islands. (After MacArthur and Wilson 1967.)

conserving species; after all, reserves are islands of suitable habitat surrounded by seas of unsuitable terrain. Much of the development of this line of reasoning came from theoretical ecologists, but some of the observations and ideas came from workers with ties to museums (e.g., Diamond 1973; Wilson and Willis 1975).

One interesting and controversial claim that arose out of these generalizations was the question of whether a single large or several small refuges would be more efficacious at preserving numbers of species. This came to be known as the "SLOSS" problem (Shafer 1990). That is, if one considered the general equations describing the behavior of island area and species numbers, it appeared that in some cases more species might be preserved in, say, ten one-hectare reserves than in one ten-hectare reserve. This line of reasoning has recently been challenged.

Patterson (1987; Patterson and Atmar 1986), a museum mammalogist, has marshaled empirical data indicating that the assemblages of species on habitat islands are not random assortments of available species but rather consist of nested subsets of species. That is, all the smaller islands of a given area will

Table 9.1 Patterns of Occurrence (+) of Native Birds on New Zealand Landbridge Islands

Island*	A	B	C	D	F	G	H	E	K	I	J	N	L	M	P	Q	S	R	V	Z	O	T	b	Y	U	a	W	X	d	e	g	h	i	j	k	c	f	m	n	o	p	s	l	r	q	t	u	4	2	v	5	z	3
1	+	+	+	+	+	+	+	+	+	+	+	+	+	+	+	+	+	+	+	+	+	+	+	+	+	+	+	+	+	+	+	+	+	+	+	+	+	+	+	+	+	+	+	+	+	+	+	+	+	+	+	+	+
2	+	+	+	+	+	+	+	+	+	+	+	+	+	+	+	+	+	+	+	+	+	+	+	+	+	+	+	+	+	+	+	+	+	+	+	+	+	+	+	+	+	+	+	+	+	+	+	+	+	+	+		+
3	+	+	+	+	+	+	+	+	+	+	+	+	+	+	+	+	+	+	+	+	+	+	+	+	+	+		+	+	+	+	+	+	+	+	+	+														+	
5	+	+	+	+	+	+	+	+	+	+	+	+	+	+	+	+	+	+	+	+	+	+	+	+	+	+	+	+	+	+	+	+	+	+	+	+																
4	+	+	+	+	+	+	+	+	+	+	+	+	+	+	+	+	+	+	+	+	+	+						+					+			+		+			+												
19	+	+	+	+	+	+	+	+	+	+	+	+	+	+	+	+	+	+	+	+	+	+	+	+	+	+						+								+		+							+				
6	+	+	+	+	+	+	+	+	+	+	+	+	+	+	+	+	+	+	+	+	+			+	+	+		+	+																								
9	+	+	+	+	+	+	+	+	+	+	+	+	+	+	+	+	+	+	+	+	+					+																											
8	+	+	+	+	+	+	+	+	+	+	+	+	+	+	+	+	+	+	+	+						+	+				+																						
12	+	+	+	+	+	+	+	+	+	+	+	+	+	+	+	+	+	+		+																																	
15	+	+	+	+	+	+	+	+	+	+	+	+	+	+	+	+	+	+	+																																		
11	+	+	+	+	+	+	+	+	+	+	+	+	+	+	+	+	+	+										+																									
10	+	+	+	+	+	+	+	+	+	+	+	+	+	+	+	+	+													+																							
7	+	+	+	+	+	+	+	+	+	+	+	+	+	+								+																															
14	+	+	+	+	+	+	+	+	+	+	+	+	+		+	+																																					
22	+	+	+	+	+	+	+	+	+	+		+	+	+	+	+							+																														
20	+	+	+	+	+	+	+	+		+	+								+		+		+																														
17	+	+	+	+	+		+																																														
13	+	+	+	+	+	+	+	+															+				+	+																									
16	+	+	+	+	+	+					+																+	+																									
18	+	+	+	+	+	+																					+	+																									
21	+	+				+						+															+																										

*See Patterson (1987: figure 2 and table 1) for key to islands and species.

have the same complement of species, and the larger islands will have these same species plus additional ones, etc. For example, Patterson showed that assemblages of birds on the various landbridge islands of New Zealand are highly nonrandom (table 9.1). Thus, in this system, species p, s, l, r, t, and u are present only on islands that have species Z, O, T, and b; these in turn are present only on islands with species A, B, C, and D. Such patterns of nested subsets show the diagonal pattern of occurrence present in table 9.1 if species are ordered by commonness. Hence, by actually examining the distribution of individual species on habitat islands, Patterson and later workers (e.g., Bolger, Alberts, and Soulé 1991) demonstrated a serious problem with island biogeography theory as it has been applied to conservation biology. If species are not interchangeable balls in an urn, then several small refuges may preserve only a subset of the taxa that might be maintained in a single larger unit.

Nevertheless, several of the generalizations arising from island biogeography do seem to hold for attempts at in-situ preservation of diversity—larger refuges are preferable to smaller ones, and corridors connecting otherwise disjunct patches are highly desirable. An experiment to test some of these predictions is currently in progress under the jurisdiction of the Smithsonian Institution (Lovejoy et al. 1986). Of course, these general guidelines are not likely to be of much concrete help in designing a plan to preserve some specific population with its own ecologic requirements.

Microconservation Biology

The general problem of designing management programs to ensure the long-term maintenance of specific populations of single taxa has become known as "population viability analysis" (Soulé 1987). It is customary to view the problem of extinction as a stochastic process, and consequently, the goals of such programs are usually couched in language such as providing a plan that will result in a 95% probability of the population not going extinct in 500 years, i.e., probability statements.

A useful way of organizing one's thoughts about the viability problem is to consider that populations may become extinct for four reasons (Shaffer 1981). First, the dynamics of very small populations may result in extinction due to lack of mates, distorted age structures, etc. That is, the demographics of populations are such that the probability of extinction increases as population size decreases; this is not a deep observation. Second, the probability of extinction also increases as population size decreases for genetic reasons; inbreeding and the expression of recessive lethals are largely responsible for this effect. Third, environmental stochasticity may also result in extinction; that is, populations of predators, parasites, and competitors will randomly fluctuate and by chance may eliminate a population. The magnitude of this effect is also an inverse function of population size. Finally, a population may become extinct as the result of natural catastrophes with effects on a geographic scale as large as or larger than that occupied by the population, e.g., drought, fires, hurricanes, floods. Obviously, there may be interactions among these four causes.

The underlying idea in this approach to the conservation of populations is that species ultimately go extinct because the last population is lost. In turn, populations disappear for only a few general reasons, enumerated above, and these factors largely reflect population sizes and their spatial geometry. This is the subject matter of population ecology and genetics, and theoreticians in these research areas have made most of the contributions to this aspect of conservation biology (e.g., Soulé 1987; Pimm, Jones, and Diamond 1988). Some of the results are summarized schematically in figure 9.3. Birth-death models of population demography indicate that the risk of extinction decreases rapidly with increasing population size. For other causes, such as short-term climatic fluctuations, environmental stochasticity, and large scale catastrophes, that may eliminate all the populations in a region, the probability of extinction decreases with increasing population size also but at slower rates.

Like demography, genetic variability scales with population size. Although it is a misleading simplification to refer to genetic variation as a single entity, it is nevertheless true that because of the monotonic scaling, the risk of extinction due to lack of genetic variation or to inbreeding should decrease rapidly as population size increases. Some systematists have been involved in these investigations as a consequence of a recent general interest in the extent of genetic variation in natural populations.

Systematists and evolutionary geneticists have devoted much time in the past two decades to documenting the quantity of genetic variation within and among conspecific natural populations (e.g., Barrowclough 1985; Hillis and Moritz 1990). This has led to a general understanding of the "normal" range of such variation and to the realization that genetic variation can easily be assayed and monitored. Thus, studies of genetic variation are now a routine part of management programs for conservation. Protein electrophoresis, for example, is widely used for this purpose. If anything, the emphasis on genetic variability has been overdone; the ease of the techniques and the availability of widely used summary statistics have led to the dominance of this aspect of viability analysis at the expense of concern for demography, environmental fluctuations, and catastrophes.

The burst of empirical results on genetic variation in natural populations also led to renewed interest in the theoretical population genetics of gene flow, mutation, selection, and population geography. The major concept in this field is the abstraction of effective population size (e.g., Crow and Kimura 1970). It was immediately realized that with regard to conservation genetics, this

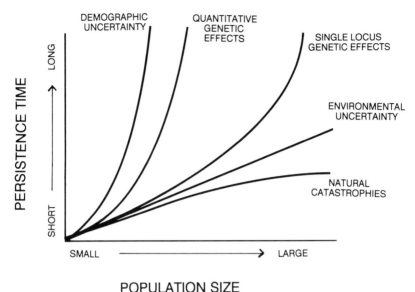

POPULATION SIZE

Figure 9.3 Relative persistence time of a population as a function of population size in the face of a variety of environmental challenges. The threat of extinction decreases rapidly with increasing population size for demographic and quantitative genetic uncertainty but less rapidly for single-locus genes. Persistence is approximately linearly related to environmental stochasticity but increases only as the logarithm of size for natural catastrophes. (After Lande and Barrowclough 1987; Shaffer 1987.)

quantity, rather than census number, was of primary importance in considering the maintenance of genetic variation in managed populations (Franklin 1980). Consequently, a large part of conservation genetics, as well as population viability analysis in general, consists of estimating the effective size of populations, assaying genetic variability, and computing how large a refuge must be to preserve an adequate effective size to prevent long-term genetic deterioration of the population. Various workers have tackled this theoretical problem over the past decade, but the most nearly comprehensive treatment is due to Lande and Barrowclough (1987), a quantitative geneticist and an avian systematist, respectively.

In the conservation literature, these workers first pointed out that it was necessary to consider separately single-locus and quantitative genetic variability and that selectively important and neutral traits would have differing behavior and requirements (e.g., table 9.2). In addition, they outlined results describing how to estimate the effective size of a population, determine the recovery time of genetic variation following severe reductions in population

Table 9.2 Effective Population Sizes Necessary for Mutation to Maintain Significant Quantities of Genetic Variation at Equilibrium. (After Lande and Barrowclough 1987; Reprinted Courtesy Cambridge University Press.)

Type of Variation	Nature of Selection	Necessary Effective Population Size	Recovery Time (Generations)
Quantitative	Neutral	~500	10^2-10^3
Quantitative	Stabilizing, or fluctuating optimum	~500	10^2-10^3
Single-locus	Neutral	10^5-10^6	10^5-10^7
Single-locus	Deleterious (incompletely recessive)	Independent of N^e (always present unless inbred)	$\sim10^2$

size, assess the influence of varying geographical structures on genetic variation, and create a monitoring protocol for the various kinds of genetic variation.

Thus, at present the theory and details of microconservation—the management of viable populations of single species—are fairly well developed from the point of view of genetics. However, the theory and necessary empirical results for coping with demographics, especially stochastic theory, environmental fluctuations, and catastrophes, as well as interactions of these aspects, are less well developed and in need of major research effort (e.g., Simberloff 1988; Soulé and Kohm 1989); the results shown in figure 9.3, for example, are only qualitatively known. Again, this is the domain of population ecology; systematists are not likely to play a major role in this research.

The Placement of Reserves

Consider the problem of where to set aside areas for reserves at the scale of biomes. For example, suppose one wished to preserve the diversity present in the Amazon basin, an area that spans nine countries. For the most part, the various taxa in this lowland forest have restricted distributions; hence several reserves would be necessary to preserve any substantial fraction of the total diversity. The placement of these reserves will clearly be a political and economic issue affecting local interests and livelihoods and influencing subsequent development. The cost of acquiring the necessary land might vary geographically. While ecologists would have much to say concerning the shape and size of the individual elements within this reserve complex, systematists are uniquely qualified to identify the general location of the reserves (Soulé 1990).

An avian systematist, Haffer (1974), was responsible for much of the modern development of this line of research. In examining patterns of avian species distribution in tropical South America, he identified regions of high species diversity, which he interpreted to be refuges of tropical vegetation during Pleistocene glaciations (figure 9.4). Many of these areas are not in the Amazon basin but occur at its edge. If these regions correspond to areas of high endemism or are potential refuges given future global climatic changes, then their study and identification are crucial for conservation schemes. Some of Haffer's interpretation has subsequently been challenged (e.g., Endler 1982), but the original observation, along with the development of the field of vicariance biogeography, led other museum and herbarium workers to try to identify what are now known as areas of endemism (e.g., Cracraft 1983). Such research, the current interest in conservation biology, and the crisis in the tropics have recently resulted in explicit attempts to identify areas of diversity and endemism in the Amazon basin that are common to such disparate taxa as mammals, birds, reptiles, amphibians, fish, insects, and plants (Rylands 1990). This is, of course, precisely what vicariance biogeography is all about (Nelson and Platnick 1981).

The integration of ideas about areas of endemism and area cladograms into the theory, literature, and practice of conservation biology has barely begun. It is another case in which the results of systematic research have received little notice but are of primary importance. This is unfortunate, for these concepts potentially provide an expanded perspective on the general problem of biological reserves that complements the current active research on the details of the management of single species and of the design of individual refuges (Shafer 1990). They also provide an evolutionary perspective on how to proceed when it becomes necessary to make difficult, irreversible decisions—the problem of triage.

Triage

Someday, the exigencies of money, human population pressure with its need for room and natural resources, and the limited space and talent for captive maintenance in zoological gardens will result in the necessity to choose among alternatives (Conway 1988). Effort will have to be devoted to one species rather than another and to one area at the expense of some other region. Such choices can be made on the basis of economics; however, alternative criteria ought to be considered also. For example, an ecologist might prefer to devote effort to maintaining a "keystone predator" rather than a producer in hopes of preserving "community structure" (Western et al. 1989; Orians and Kunin 1990; Soulé 1990). Other viewpoints ultimately involving aesthetics

Figure 9.4 Presumed locations of forest refugia in tropical South and Central America during arid climatic periods of the Pleistocene. (Haffer 1974: figure 13.1; reprinted courtesy of the Nuttall Ornithological Club.)

could be invoked; for example, charismatic taxa might generate popular support not available for parasites.

Given that difficult decisions will have to be made, systematists may have unique perspectives on the relative importance of taxa and areas at variance with current orthodoxy. For example, at present, two views (not alternates) on the biodiversity crisis are widespread. First, there is the goal of saving all species and perhaps all genetically distinctive populations (Ryder 1986). This is admirable but ultimately unlikely to be realized. For example, will the philosophy extend to all of the millions of species of insects? In truth, this view puts off troubling analysis and discussion into the indefinite future. A second view commonly voiced is the concern about the tropics and "rainforests," with their "great diversity" (Myers 1988; Lean, Hinrichsen, and Markham 1990). An unstated assumption in this concern is that diversity is equivalent to species number and that extrapolation from numbers of species of trees and birds to all other taxa is reasonable. That the latter is not true in general is clear from Platnick's (this volume) observations on the worldwide distribution of spiders. More importantly, however, the implicit assumption that diversity equals numbers ignores everything that has been learned about systematic information and hierarchical relationships.

The possibility that all species should not be treated equally in conservation triage, based on their hierarchical relationships, entails a systematic and evolutionary viewpoint not widely appreciated. Vane-Wright, Humphries, and Williams (1991) and Stiassny (this volume) have discussed this in a "quantity vs. quality" framework. As a simple example, consider the problem of the relative value of the two species of tuatara, which are the sister group to the other 6800 species of snakes and lizards; most taxonomists would recognize the need to give special consideration to these New Zealand taxa, in part because of the information they provide about early divergences (May 1990). Similarly, Stiassny (this volume) has pointed out that some nine Malagasy fish are sister to all other cichlids, including the famous radiations in the east African lakes. As a final example, consider that the results of several ornithological studies (e.g., Cracraft and Prum 1988) suggest that some of the taxa in the coastal forest of southeastern Brazil have a sister group relationship to other species throughout the Amazon basin and hence are in some sense equivalent to that entire avifauna. Thus, if one were to set priority for conservation activities based on systematic content, the distribution of effort might not conform to widespread preconceptions.

Museum researchers have been at the forefront of activity aimed at developing a quantitative measure of the "systematic value" of taxa. For example, Vane-Wright, Humphries, and Williams (1991) have proposed an index that evaluates taxa based on their phylogenetic position; relatively older taxa and members of less speciose clades receive more weight in allocating limited preservation effort. Consider, for example, figure 9.5. A species of parrot inhabiting southeastern Brazil, *Pionopsitta pileata,* is known to be the sister group to the rest of the genus. In the Vane-Wright, Humphries, and Williams weighting scheme, this species would receive more conservation effort than the combined three species in the Amazon basin.

Analyses such as this are in their infancy and require additional refinement (e.g., May 1990); however, the underlying philosophy and novel viewpoint should generate useful controversy and lead to a reassessment of the ideas and goals in the biodiversity effort.

The biodiversity crisis involves local, biomic, and global geographic scales. Of primary importance to the practicing systematist is the intermediate level because at this scale conservation biology operates in documenting and preserving species; here the talents of museum workers are essential.

Within the field of conservation biology, systematists have made many and diverse contributions; however, their role is primary in the documentation of diversity, including the description of undescribed species, the characterization of patterns of intraspecific variation, and the ordering of information about taxonomic hierarchies into useful generalizations about geographic relationships, etc.

Necessary but secondary roles for systematic researchers include communicating the importance of documenting biodiversity in order to obtain an understanding of evolutionary history and natural processes; this view should complement the opinions of nonhistorical specialists such as economists, ecologists, and politicians about where and how to allocate resources.

With regard to the actual mechanics of preservation, systematists have made important contributions to both macroconservation and microconservation theory. The former involves developing phenomenological predictions over ensembles of taxa, in the absence of extensive data about individual species and their needs. In this research, observations by taxonomists about biogeography and distributions have led to generalizations used in designing refuges for species preservation. Microconservation biology concerns the development of methodology for preserving single species based on specific ecologic and genetic requirements. Systematists have been involved in the latter because of widespread interest in recent years in documenting patterns of genetic variation in natural populations. Other aspects of single-species management have largely been the domain of population ecologists.

ACKNOWLEDGMENTS

I would like to thank Niles Eldredge, Nate Flesness, Rick Prum, Les Short, and Melanie Stiassny for useful discussions of issues raised in this paper; in addition, Niles

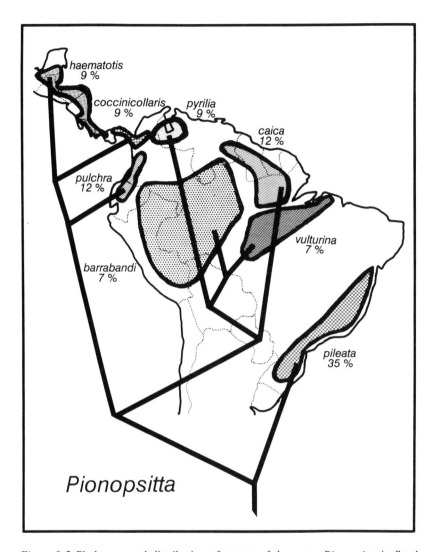

Figure 9.5 Phylogeny and distribution of parrots of the genus *Pionopsitta* in South and Central America. The relative amount of conservation effort that ought to be allocated to each species is indicated as a percentage following the taxonomic weighting scheme of Vane-Wright, Humphries, and Williams (1991). Distribution and phylogeny after Cracraft and Prum (1988).

Eldredge, Patricia Escalante, Rosemary Gnam, and Carole Griffiths provided helpful comments on the manuscript. Paul Sweet prepared several of the figures.

REFERENCES

Barrowclough, G. F. 1985. Museum collections and molecular systematics. In E. H. Miller, ed., *Museum Collections: Their Roles and Future in Biological Research,* pp. 43–54. Victoria, B.C., Canada: British Columbia Provincial Museum.

Barrowclough, G. F. and N. R. Flesness. 1992. Species, subspecies, and races: The problem of the units of management in conservation. In M. Allen and H. Harris, eds., *Wild Mammals in Captivity.* Chicago: University of Chicago Press (in press).

Barrowclough, G. F. and R. J. Gutiérrez. 1990. Genetic variation and differentiation in the Spotted Owl (*Strix occidentalis*). *Auk* 107:737–744.

Bolger, D. T., A. C. Alberts, and M. E. Soulé. 1991. Occurrence patterns of bird species in habitat fragments: Sampling, extinction, and nested species subsets. *American Naturalist* 137:155–166.

Callicott, J. B. 1986. On the intrinsic value of nonhuman species. In B. G. Norton, ed., *The Preservation of Species,* pp. 138–172. Princeton: Princeton University Press.

Capparella, A. P. 1988. Genetic variations in Neotropical birds: Implications for the speciation process. In H. Ouellet, ed., *Acta XIX Congressus Internationalis Ornithologici,* pp. 1658–1664. Ottawa, Canada: University of Ottawa Press.

Coddington, J. A. 1988. Cladistic tests of adaptational hypotheses. *Cladistics* 4:3–22.

Conway, W. 1988. Can technology aid species preservation? In E. O. Wilson, ed., *Biodiversity,* pp. 263–268. Washington, D.C.: National Academy Press.

Cracraft, J. 1983. Cladistic analysis and vicariance biogeography. *American Scientist* 71:273–281.

Cracraft, J. and R. O. Prum. 1988. Patterns and processes of diversification: Speciation and historical congruence in some Neotropical birds. *Evolution* 42:603–620.

Crow, J. F. and M. Kimura. 1970. *An Introduction to Population Genetics Theory.* New York: Harper and Row.

Diamond, J. M. 1973. Distributional ecology of New Guinea birds. *Science* 179:759–769.

Easterbrook, G. 1990. Everything you know about the environment is wrong. *New Republic* 3928:14–27.

Ehrenfeld, D. 1988. Why put a value on biodiversity? In E. O. Wilson, ed., *Biodiversity,* pp. 212–216. Washington, D.C.: National Academy Press.

Eldredge, N. and J. Cracraft. 1980. *Phylogenetic Patterns and the Evolutionary Process.* New York: Columbia University Press.

Endler, J. A. 1977. *Geographic Variation, Speciation, and Clines.* Princeton: Princeton University Press.

Endler, J. A. 1982. Pleistocene forest refuges: Fact or fancy? In G. T. Prance, ed., *Biological Diversification in the Tropics,* pp. 641–657. New York: Columbia University Press.

Erwin, T. L. 1988. The tropical forest canopy: The heart of biotic diversity. In E. O. Wilson, ed., *Biodiversity*, pp. 123–129. Washington, D.C.: National Academy Press.

Escalante-Pliego, B. P. 1992. Genetic differentiation in yellowthroats (Parulinae: *Geothlypis*). In B. D. Bell, ed., *Acta XX Congressus Internationalis Ornithologici*. Wellington, N.Z.: New Zealand Ornithological Congress (in press).

Farnsworth, N. R. 1988. Screening plants for new medicines. In E. O. Wilson, ed., *Biodiversity*, pp. 83–97. Washington, D.C.: National Academy Press.

Felsenstein, J. 1985. Phylogenies and the comparative method. *American Naturalist* 125:1–15.

Franklin, I. R. 1980. Evolutionary change in small populations. In M. E. Soulé and B. A. Wilcox, eds., *Conservation Biology: An Evolutionary-Ecological Perspective*, pp. 135–149. Sunderland, Mass.: Sinauer Associates.

Freedman, B. 1989. *Environmental Ecology*. San Diego: Academic Press.

Funk, V. A. and D. R. Brooks. 1990. Phylogenetic systematics as the basis of comparative biology. *Smithsonian Contributions to Botany* 73:1–45.

Haffer, J. 1974. *Avian Speciation in Tropical South America*. Cambridge, Mass.: Nuttall Ornithological Club.

Highton, R., G. C. Maha, and L. R. Maxson. 1989. Biochemical evolution in the slimy salamanders of the *Plethodon glutinosus* complex in the eastern United States. *Illinois Biological Monographs* 57:1–153.

Hillis, D. M. 1988. Systematics of the *Rana pipiens* complex: Puzzle and paradigm. *Annual Review of Ecology and Systematics* 19:39–63.

Hillis, D. M. and C. Moritz. 1990. *Molecular Systematics*. Sunderland, Mass.: Sinauer Associates.

Iltis, H. H. 1988. Serendipity in the exploration of biodiversity: What good are weedy tomatoes? In E. O. Wilson, ed., *Biodiversity*, pp. 98–105. Washington, D.C.: National Academy Press.

Lande, R. and G. F. Barrowclough. 1987. Effective population size, genetic variation, and their use in population management. In M. E. Soulé, ed., *Viable Populations for Conservation*, pp. 87–123. Cambridge, U.K.: Cambridge University Press.

Lean, G., D. Hinrichsen, and A. Markham. 1990. *Atlas of the Environment*. New York: Prentice Hall.

Lovejoy, T. E. 1986. Species leave the Ark one by one. In B. G. Norton, ed., *The Preservation of Species*, pp. 13–27. Princeton: Princeton University Press.

Lovejoy, T. E., R. O. Bierregaard, Jr., A. B. Rylands, J. R. Malcolm, C. E. Quintela, L. H. Harper, K. S. Brown, Jr., A. H. Powell, G. V. N. Powell, H. O. R. Schubart, and M. B. Hays. 1986. Edge and other effects of isolation on Amazon forest fragments. In M. E. Soulé, ed., *Conservation Biology: The Science of Scarcity and Diversity*, pp. 257–285. Sunderland, Mass.: Sinauer Associates.

MacArthur, R. H. and E. O. Wilson. 1967. *The Theory of Island Biogeography*. Princeton: Princeton University Press.

May, R. M. 1990. Taxonomy as destiny. *Nature* 347:129–130.

Myers, N. 1988. Tropical forests and their species: Going, going . . . ? In E. O. Wilson, ed., *Biodiversity*, pp. 28–35. Washington, D.C.: National Academy Press.

Myers, P. and J. L. Patton. 1989. A new species of *Akodon* from the cloud forests of eastern Cochabamba Department, Bolivia (Rodentia: Sigmodontinae). *Occasional Papers of the Museum of Zoology, University of Michigan* 720:1–28.

Nelson, G. and N. Platnick. 1981. *Systematics and Biogeography: Cladistics and Vicariance.* New York: Columbia University Press.

Norton, B. G. 1987. *Why Preserve Natural Variety?* Princeton: Princeton University Press.

Oren, D. C. and H. Guerreiro de Albuquerque. 1991. Priority areas for new avian collections in Brazilian Amazonia. *Goeldiana Zoologia* 6:1–11.

Orians, G. H. and W. E. Kunin. 1990. Ecological uniqueness and loss of species. In G. H. Orians, G. M. Brown, Jr., W. E. Kunin, and J. E. Swierzbinski, eds., *The Preservation and Valuation of Biological Resources,* pp. 146–184. Seattle, Wash.: University of Washington Press.

Patterson, B. D. 1987. The principle of nested subsets and its implications for biological conservation. *Conservation Biology* 1:323–334.

Patterson, B. D. and W. Atmar. 1986. Nested subsets and the structure of insular mammalian faunas and archipelagos. *Biological Journal Linnean Society* 28: 65–82.

Pimm, S. L., H. L. Jones, and J. Diamond. 1988. On the risk of extinction. *American Naturalist* 132:757–785.

Pinedo-Vasquez, M., D. Zarin, P. Jipp, and J. Chota-Inuma. 1990. Use-values of tree species in a communal forest reserve in northeast Peru. *Conservation Biology* 4:405–416.

Platnick, N. I. 1992. In this volume.

Plotkin, M. J. 1988. The outlook for new agricultural and industrial products from the tropics. In E. O. Wilson, ed., *Biodiversity,* pp. 106–116. Washington, D.C.: National Academy Press.

Radulovich, R. 1990. A view on tropical deforestation. *Nature* 346:214.

Risebrough, R. W. 1986. Pesticides and bird populations. *Current Ornithology* 3: 397–427.

Ryder, O. A. 1986. Species conservation and systematics: The dilemma of subspecies. *Trends in Ecology and Evolution* 1:9–10.

Rylands, A. B. 1990. Priority areas for conservation in the Amazon. *Trends in Ecology and Evolution* 5:240–241.

Shafer, C. L. 1990. *Nature Reserves: Island Theory and Conservation Practice.* Washington, D.C.: Smithsonian Institution Press.

Shaffer, M. L. 1981. Minimum population sizes for species conservation. *Bioscience* 31:131–134.

Shaffer, M. 1987. Minimum viable populations: Coping with uncertainty. In M. E. Soulé, ed., *Viable Populations for Conservation,* pp. 69–86. Cambridge, U.K.: Cambridge University Press.

Silver, C. S. 1990. *One Earth, One Future: Our Changing Global Environment.* Washington, D.C.: National Academy Press.

Simberloff, D. 1988. The contribution of population and community biology to conservation science. *Annual Review of Ecology and Systematics* 19:473–511.

Soulé, M. E. 1987. *Viable Populations for Conservation.* Cambridge, U.K.: Cambridge University Press.

Soulé, M. E. 1990. The real work of systematics. *Annals of the Missouri Botanical Garden* 77:4–12.

Soulé, M. E. and K. A. Kohm. 1989. *Research Priorities for Conservation Biology.* Washington, D.C.: Island Press.

Stuessy, T. F. and K. S. Thomson. 1981. *Trends, Priorities and Needs in Systematic Biology.* Lawrence, Kansas: Association of Systematic Collections.

Stiassny, M. L. J. 1992. In this volume.

Sullivan, A. L. and M. L. Shaffer. 1975. Biogeography of the megazoo. *Science* 189:13–17.

Vane-Wright, R. I., C. J. Humphries, and P. H. Williams. 1991. What to protect? Systematics and the agony of choice. *Biological Conservation* 55:235–254.

Western, D., M. C. Pearl, S. L. Pimm, B. Walker, I. Atkinson, and D. S. Woodruff. 1989. An agenda for conservation action. In D. Western and M. C. Pearl, eds., *Conservation for the Twenty-First Century,* pp. 304–323. New York: Oxford University Press.

Wheeler, Q. D. 1989. Militant view of needs and priorities for training systematic biologists. *Association of Systematics Collections Newsletter* 17:45–52.

Wilson, E. O. 1988a. *Biodiversity.* Washington, D.C.: National Academy Press.

Wilson, E. O. 1988b. The current state of biological diversity. In E. O. Wilson, ed., *Biodiversity,* pp. 3–18. Washington, D.C.: National Academy Press.

Wilson, E. O. and E. O. Willis. 1975. Applied biogeography. In M. L. Cody and J. M. Diamond eds., *Ecology and Evolution of Communities,* pp. 522–534. Cambridge, Mass.: Harvard University Press.

Zink, R. M. and D. L. Dittmann. 1991. Evolution of Brown Towhees: Mitrochondrial DNA evidence. *Condor* 93:98–105.

10 : Systematics and Marine Conservation

Judith E. Winston

The State of the Oceans

San Diego Bay. Changes in hard-bottom communities are blamed on tributyltin, a pesticide used in antifouling paints (Lenihan, Oliver, and Stephenson 1990). This chemical is known to destroy gastropod populations by causing female sex changes that result in sterility (Gibbs, Pascol, and Burt 1988, Spence, Hawkins, and Santos 1990). Other studies substantiate its high toxicity to bivalve mollusks and invertebrate larvae (Beaumont et al. 1989).

Pacific Coasts of Panama and Costa Rica. Dinoflagellate bloom causes death of reef corals (Guzman et al. 1990).

Caribbean. Widespread outbreaks of damaging "bleaching" (loss of symbiotic algae) are reported (Roberts 1990).

Baja California Sur, Mexico. From 50% to 95% of *Pocillopora* reef corals are dead a year after a bleaching event was reported for the area (Wilson 1990).

Gulf of Mexico. Substantial amounts of hydrocarbon- and barium-rich sediment and pit sludge, plus increasing concentrations of biocides, are reported from oil- and gas-producing areas (Fang 1990).

Southeast Coast, United States. Dying dolphins wash up on Atlantic beaches from New Jersey to Florida. Scientists discover that their deaths are due to brevetoxin poisoning caused by ingesting menhaden that had fed upon a toxic red tide alga (Joyce 1989).

New Jersey Coast. Stomach contents of a 150 lb bluefin tuna caught by recreational fishermen include the following items: cocaine inhalation straw, elastic pony tail holder, candy wrappers, elastic bands, nylon fishing line, balloons, plastic bags, pens, and markers (Bergquist, Giebel, and Rudman 1990).

Cross Island, Maine. In 1987, 300 kg of drift plastic are collected from the shoreline of this uninhabited island. Just one year later another 124 kg of plastic debris are collected. Both studies show accumulation of plastic debris just above the intertidal zone—where birds and mammals may easily become entangled (Podolsky 1989).

Canada. Winter flounder gene for antifreeze protein is successfully introduced into Atlantic salmon (Fletcher, Shears, King, Davies, and Hew 1988).

Prince William Sound, Alaska. *Exxon Valdez* goes aground spilling 11 million gallons of crude oil. Immediate effects include deaths of as many as 300,000 seabirds and more than 1000 sea otters; long-term effects caused by chemical pollution or eutrophication still unknown. The most distressing results to biologists include the damage caused by cleanup attempts, the legal (and sometimes physical) battles by competing groups over the evidence (e.g., marine mammal carcasses), and the lack of support for long-term studies (Isleib 1989; Davidson 1990; Drew 1990, Michel 1990; Turner 1990).

Arthur Harbor, Antarctica. *Bahia Paraiso* goes aground spilling more than 150,000 gallons, mostly arctic diesel fuel. Immediate effects (confined to within a few km of the wreck) include deaths of some adult seabirds (penguins and shags) and many shag and skua nestlings. Timing of the event (after most penguin chicks had left the rookery) and the physical conditions (steep rocky shores and strong wave and wind action) limit damage, but long-term sublethal effects are still possible (Kennicut et al. 1990).

Antarctica. Photo of a dying fur seal entangled in plastic packing bands shows that the effects of plastic debris reach even this far (photo by Bonner in *IUCN Bulletin,* 21(2):20).

Sydney, Australia. During the height of the tourist season, Bondi Beach, one of Australia's chief attractions, was declared unsafe for swimming two days out of five due to sewage pollution. Pollution and contamination have damaged the local fishing industry as well. In spite of protests, effective cleanup action is not imminent (Beder 1990; Priestley 1990).

South Africa. A scientific assessment of southern African estuaries shows only 24% to 28% could be considered to be in good ecological condition (Heydorn 1989).

European Coasts. Seal plague causes 18,000 deaths of Harbor seals. Although illness was probably caused by viral infection, pollution may have weakened seals' immune systems (Dietz, Heide-Jorgensen, and Harkonen 1989; Harwood 1989).

Mediterranean. Mass deaths of dolphins are likewise linked to deadly effects of epidemic virus and pollution. Autopsies show high levels of toxic chemicals, including heavy metals and PCBs (Simons 1990).

Scandinavian Coastal Waters, North Sea. Changes in species composition, a decline in seaweed abundance, and deaths of benthic invertebrates and fish are attributed to a series of toxic algal blooms (Rosenberg, Lindahl, and Blanck 1988; Underdal et al. 1989; Baden, Loo, Pihl, and Rosenberg 1990; Pedersen, Björk, Larsson, and Söderlund 1990).

The Hague, Netherlands. Representatives of nine European governments reach agreement on a number of measures to cut pollution of the North Sea ecosystem by heavy metals, pesticides, and other chemical compounds. Meanwhile Dutch scientists argue that overfishing is also destructive and call for a ban on fishing in a quarter of its area (Milne 1990).

World Oceans. Fishery statistics indicate that the world harvest of fish has leveled off or started to decline owing to habitat loss and overfishing (Bennett 1990).

Baltic Sea. Meanwhile, Baltic fish communities have changed in composition, but total fish catches have doubled in the last 25 years, owing to eutrophication and decreased predation by seals (whose populations have greatly decreased owing to hunting and pollution) (Hansson and Rudstam 1990).

Arabian Gulf, Indian Ocean. Actions of recent war released largest oil spill ever (more than 10 million *barrels*) into waters already stressed by high temperatures, oil pollution, dredging, and landfill. Vast shallow-water seagrass and algal ecosystems, which support shrimp and juvenile pearl oysters, migrating seabirds, and nesting sea-turtles, are immediately affected (Raloff and Monastersky 1991; Sheppard and Price 1991).

World Oceans. Scientists predict spread of coral reefs with rising sea level. While growing reefs lock up carbon in coral skeletons (discouraging the greenhouse effect) they may contribute more carbon to the atmosphere via carbon dioxide production, thus encouraging the greenhouse effect (Pain 1990).

Earth is a marine planet—salt water covers about 70% of its surface. Because the entire water column, as well as the sea bottom and the surface, is inhabited, the oceans are even more important in terms of living space—they contain 99% of the planet's supply (Barnes and Hughes 1982).

The seas are our original home. In them, more than 3 billion years ago, life on earth began. Only much more recently (about 400 million years ago) did animals and plants take up life on land.

Human beings are among those successful terrestrial organisms, but the marine realm remains important to our lives. Human populations are concentrated near the oceans' margins—70% of all people live along the earth's coastal plains.

These marginal environments—coasts, bays, and estuaries—are the richest part of the oceans. They account for 30% of the oceans' productivity, even though they encompass only 10% of its surface and a tiny fraction (0.5%) of its volume (Cherfas 1990). Yet we continue to treat these essential resources with little respect. Coastal environments, especially estuaries, have been treated as "free sewers for coastal cities" (Odum 1989), even though most of the seafood we eat spends at least its early stages in estuarine nursery grounds. And, as the above examples show, our waste products and pollutants are increasingly affecting not only the estuaries and harbors but also the open coasts and even the continental shelves.

The incidents listed above are just a few of the most recent, and the problems are continual and increasing. They fall into several categories:

1. *habitat loss,* e.g., destruction of wetland, marshes, and mangroves, and damming of freshwater inflow, with resultant increases or fluctuations in salinity
2. *eutrophication,* excess nutrient input—what a marine biologist from Germany recently labeled our "green hand in the sea" (Smetacek in Cherfas 1990)
3. *chemical pollution*
4. *introductions of exotic species*—by a conservative estimate more than a thousand species have had their distributions altered by human activity (Carlton 1989)
5. *direct genetic changes* caused by biotechnology: manipulation of the genes of commercially valuable marine species
6. *indirect genetic changes* caused by overexploitation of marine species, which may exert enough selection pressure to actually affect a species' evolution (Law 1991)
7. *global changes* that could overturn the entire circulation pattern of the oceans, an event whose catastrophic results would make the first six categories irrelevant (e.g., Wilson 1989; Holligan 1990; Williamson and Holligan 1990; Sathyendranath et al. 1991).

Diversity of the Oceans

Terrestrial environments have the greatest diversity in terms of total numbers of known species (chiefly because of the insects and other arthropods), but in terms of phylogenetic diversity—basic types of body architecture—and in terms of life cycles and habitats, the greatest diversity is in the sea (Ray 1988).

Biologists recognize five major categories of organisms or Kingdoms: Monera, including blue-green algae and bacteria; Protista; Fungi; Plantae (metaphyte plants); and Animalia (metazoan animals). All have marine representatives. Of the major groupings within them (phyla), all but a few are represented in the sea. There are about 16 phyla of monerans (Lane 1990) and 67 phyla of eukaryotic organisms (Barnes 1984).

But additional, still undiscovered phyla may exist in the sea. The most recently discovered marine phylum—the Loricifera—was described only in 1983 (Kristensen 1983). Several new classes of marine invertebrates have been discovered in recent years, including the Concentricycloidea (echinoderms) and the Remipedia, Tantocarida, and Cephalocarida (crustaceans) (Barnes 1984).

These recent discoveries of major groups of marine animals indicate that our knowledge of marine diversity is far from complete. This is partly due to the technical and logistical problems of sampling marine environments. Systematic sampling of the deeper parts of the ocean did not begin until the first half of the nineteenth century (Soule 1988). Even now, at the end of the twentieth century, while our destructive influence reaches further and further out into the sea, our ability to sample its flora and fauna is still largely reliant on nineteenth-century techniques. Some of the greatest advances in marine biology and ecology have come through the general scientific use of SCUBA diving techniques during the last twenty-five years. But one has only to dive to compressed air SCUBA limits (about 150 feet) on a Caribbean reef wall and look down into the blue abyss to realize what a small portion of the marine realm our present knowledge encompasses. Even today most marine research is limited to the intertidal or shallow subtidal zones, to the coasts and the inner part of the continental shelves, to what we can collect by diving or by using dredges, trawls, and grabs not much different than those employed by *Challenger* Expedition scientists in the 1870s.

Conventional methods like fishing and dredging have informed us that animals thought to be extinct—the coelocanth, the monoplacophoran mollusks, the megamouth shark—still inhabit the ocean depths. It is not unrealistic to believe that other animals, even very large animals, could still be hidden there (Barnes and Hughes 1982; Leblond 1990).

Where new methods have been invented and implemented, the results have been astonishing. For example, plankton used to be sampled with nets. Even the finest plankton nets allowed all organisms less than 20 μm in size to pass through. New sampling techniques like fluorescence microscopy of water samples and direct plate counts of bacteria have shown us that these smallest members of the plankton—the nanoplankton (2–20 μm, including both photosynthetic and heterotrophic forms from a variety of protistan groups) and the picoplankton (0.2–2 μm, including a few chlorophytes, as well as bacteria and cyanobacteria)—are very important to productivity in different areas of the ocean. Whole new groups of algae have been discovered this way, including some whose systematic positions are still unknown. We now realize that many of these smallest elements of the plankton can exist deeper in the ocean than was ever thought possible.

The discovery of this "microbial loop" in the pelagic food chain has made all old estimates of primary productivity underestimates. But as a recent reviewer of the subject has put it, "further progress towards a full understanding of the structure and diversity of planktonic communities will, however, depend on an increased knowledge about individual species" (Fenchel 1988).

Ideally, manned submarines and remotely operated vehicles (ROVs) could be used to extend our knowledge into the deeper parts of the oceans. Their use to study pelagic ecosystems could enable us to more fully understand and predict the oceans' biological productivity, as well as the distribution and diversity of benthic communities. For example, during the last ten years, scientists using submersibles equipped with robotic collecting arms have captured and photographed marvelous gelatinous zooplankton, previously either unknown or known only from remnants of colonies scrambled into jelly by collecting nets (Fenaux and Youngbluth 1990; Youngbluth, Bailey, and Jacoby 1990). Other scientists, studying echninoderms from the same vehicles, have observed alive, videotaped, and collected intact deep-water species either new to science or previously known only from a few broken and fragmentary dredged specimens (Miller and Pawson 1989).

State of Taxonomic Knowledge of Marine Groups

It is clear that many new marine species and even new higher taxa still remain to be described. But even our knowledge of the groups we know exist in the ocean is far from complete. Because compilations of numbers of described species are usually listed only taxonomically and not by habitat, it is difficult even to discover how many species have been described from the oceans. According to the recently published *Atlas of the Environment* (Lean, Hinrichsen, and Markham 1990), about 1.4 million species have been identified from all environments on earth. According to Barnes and Hughes (1982) about 250,000 of these are marine. In table 10.1 I have listed total numbers of marine species described by group, based on various sources, and have come up with a similar figure.

But how well does the described diversity reflect the true diversity of the oceans—what percentage of species still remains to be described? Obviously, there is no way to figure this out directly, but there are several indirect approaches to the problem.

One approach is to tally the numbers of new species that have been described in a group over a period of years. If a large or increasing number of species is still being described, it implies that our knowledge of the group is still small relative to its total diversity. When R. D. Barnes took that approach (1987, 1989) he found that for some groups (e.g., rotifers and cnidarians) numbers changed little over a 30-year period (1952–1982). Other groups (e.g., sponges, annelids, bryozoans, echinoderms) showed moderate increases, and a few groups (flatworms, crustaceans, and arachnids) showed large increases in species numbers. However, this method can be misleading. Change or stasis in species numbers may have occurred for reasons having nothing to do with the total numbers of species. For example, if new species were described, but old ones were deleted during the same period by being placed in synonymy, the total would remain the same and real changes in knowledge would be obscured. The apparent growth of a field is also highly dependent on the numbers of people working on a group. Most groups are so poorly represented in terms of specialists that a few deaths, or the appearance of a few enthusiastic new specialists, can bias results for that group.

Another way to answer the question is to ask the specialists themselves for their estimates of the percentage of species that they think still remain to be described in their specialty. I took this approach recently for a symposium on the biomedical importance of marine organisms, asking 20 of my colleagues to complete a questionnaire on the organisms they studied (Winston 1988). My results (table 10.2) are limited, for they included only the marine plant and animal groups most relevant to those studying natural products chemistry and only the geographic areas in which chemical defenses are most frequently encountered, but they are still instructive.

It is apparent that some groups and some areas are much better known than others. For example, hydroids seem relatively well known almost everywhere, while nemerteans are poorly known.

The coasts of the United States appear fairly well known with an overall average of about 80% of the flora and fauna in the selected groups thought to be described. The Caribbean is the best studied tropical area, with about 70% of species estimated to be known. In the non-Caribbean tropics the figure drops to 50–60% of the total flora and fauna. I had listed the Australian Great Barrier Reef (a favorite collecting ground for natural products chemists) sepa-

Table 10.1 Estimated Numbers of Species of Major Groups of Marine Organisms

Group	No.
Viruses	1,000
Eubacteria	3,000
Cyanobacteria	1,800
Protista (27 phyla)	23,000
Algae	
Chlorophyta	7,000
Phaeophyta	1,500
Rhodophyta	4,000
Marine fungi	498
Seagrasses	50
Invertebrates	
Sponges	10,000
Coelenterates	10,000
Ctenophores	90
Platyhelminthes	15,000
Nemerteans	750
Nematodes	12,000
Gnathustomulids	80
Gastrotrichs	400
Kinorhynchs	100
Acanthocephalans	600
Rotifers	200
Entroprocts	170
Phoronids	16
Brachiopods	350
Bryozoans	5,000
Mollusks	75,000
Sipunculans	250
Priapulids	8
Annelids	12,000
Echiurids	130
Crustaceans	39,000
Pycnogonids	1,000
Marine mites	300+
Chaetognaths	50
Hemichordates	100
Echinoderms	6,000
Ascidians	2,000
Cephalochordates	23
Vertebrates	
Lampreys and hagfish	25
Skates, sharks, and rays	520
Bony fish	11,675
Sea turtles	6
Sea snakes	50
Manatees	5
Cetaceans	76

Table 10.1 Continued

Group	No.
Seals, walruses, sea lions	31
Otters	1
Seabirds[a]	312
Total	245,166+

Sources: Grzimek 1968; Cohen 1970; Parker 1982; Moss 1986; Woolley 1986; Barnes 1989; Lane 1990; May 1990.
  [a]Not including shorebirds.

rately because I thought it might be better known than the rest of the Pacific, but this did not seem to be the case. I included one other nontropical area of interest to the chemists in my survey—Antarctica. Perhaps because of the interest generated in this last frontier, Antarctica has received more study—about 70% of species are estimated to be known.

Table 10.2 points out that much basic taxonomic work remains to be done on marine organisms. It also shows how unpromising the situation can be for a person (like an ecologist or a natural products chemist) seeking to identify a specimen—in many tropical areas the chances are 50/50 that the species to which it belongs has not been described. It becomes even more discouraging when you consider that a number of these groups (like mollusks) are among the best studied marine organisms.

Attempting any prediction of the total numbers of species that may remain to be found in the sea is probably foolhardy, but, basing the guess on a total number of known marine species of about 250,000 (table 10.1), we can arrive at some maximum and minimum figures. If the whole sea were as well known as the European or North American coasts (70–80%), there would still be 50,000–75,000 species to be discovered there. If it were like much of the tropic and subtropic regions (50–60%), there could be 150,000–200,000 species not yet described. If the deep sea, as well as the continental shelves and coastal margins, is considered, the total could reach half a million or more.

Should we care about such numbers? We have certainly heard enough about the need to preserve the rainforest and its plants—but *all* the known flowering plants in the world comprise only about 240,000 species (Stace 1989).

Of course, the overwhelming components of terrestrial diversity, with about 750,000 species described and between 10 and 30 million predicted (Erwin 1983; May 1988), are the insects and arachnids, many of which are tiny mites found on other species of insects.

Table 10.2 Estimates of Percent of Marine Fauna Known for Selected Groups and Areas

Group	E. Coast, USA	W. Coast, USA	Caribbean	Indo-Pacific	Great Barrier Reef	E. Pacific	Africa	Antarctica
Algae	80	90	80	60	60	70	70	80
Sponges	75	75–80	60–65	60	60	50	40	50
Corals	95	80	95	70–80	70?	70–80	70	90
Octocorals	60–70	50	60–70	50	—	50	75	50
Hydroids	98	93	95	80–85	93	87	93	90
Sea anemones	—	90+	75	75	50	—	—	95
Nemerteans	70–80	50	20	20	20	20	20	20
Bryozoans	75	70	60	50	50	40	70–80	40–50
Ascidians	80	65–70	60	55	50	25–50	25–50	95
Opisthobranch mollusks	80–90	80–90	60	20–30	40–50	40	70–80	40–50
Ophiuroids	85	90	80	60	—	—	—	70
Echinoids and holothurians	80	80	80	80	80	80	80	80

From Winston 1988, table 3

All estimates are for continental shelf depths or shallower. Figures would decline sharply for deeper water. U.S. coasts included for comparison with tropical areas, since, in view of the presence of well-established marine laboratories, they could be expected to be reasonably well studied.

Likewise, the estimate of 250,000 marine organisms leaves out the "mites" of the sea, which must make up a good proportion of marine diversity. There are real marine mites of course. Their numbers are estimated at more than 300 species (Woolley 1986), but, as with their terrestrial relatives, this probably represents only a small proportion of their total diversity. When we include them, along with all the other meiofaunal to microscopic organisms in the sea: the phytoplankton (especially the microplankton to picoplankton), the foraminiferans, ciliates and other protistan groups, as well as the varied and ubiquitous megafaunal categories, the total numbers of unknown marine species could reach a million or more.

Systematics and Marine Biodiversity

And who is there to study and describe those unknown marine hordes? Only a token number of systematists. For the same symposium I estimated the numbers of people working on chemically important groups of marine organisms in each of the geographic areas of table 10.2. This resulted in another table that had 96 combinations (each group by each area). Sixty-three combinations of the 96 contained 3 or fewer systematists (Winston 1988). Anyone who had a question concerning identification for one of those groups in 1 of those 63 areas would have 3 or fewer specialists to turn to for help.

The groups studied had a total of about 150,000 described species—but worldwide fewer than 200 people were actively pursuing systematic studies on them. Since my definition of people actively publishing on a group included some who spent only part of their time on systematics, as well as some specializing in a subgroup only (e.g., a single family, or geographic area), it greatly understates the scarcity of expertise, even on those relatively well-studied groups. To contrast the situation with that for vertebrates, there are about 50 mammal specialists in the United States alone, to cover the approximately 4,000 known mammal species. Marine groups with similar numbers of described species—like corals (4,000) and bryozoans (5,000) cannot boast half that many systematists in the entire world. Many marine groups are much less actively studied. Some have not a single living specialist.

Numerous reports over the last 25 years have documented the problems faced by the field of systematics, including rapidly growing collections, inadequate funding for equipment and curatorial supplies, and installation or upgrading of data management systems (Steere et al. 1971; Irwin et al. 1973; Stuessy and Thompson 1981; Edwards, Davis, and Nevling 1985; Scudder 1987). But although famous biologists have spoken out about our urgent need for systematics (May 1988, 1990, E. O. Wilson 1985, 1988, 1991), the situation at the start of the 1990s seems more discouraging than ever.

The world's population of systematists is an aging population, many of them in their fifties or older. Few students are going into the field. Many U.S. college and university biology departments discourage the pursuit of systematic research by graduate students and faculty. A recent report on systematics and evolutionary biology in Great Britain shows a similar trend there. British museums have lost 20–25% and universities at least 35% of their evolutionary positions in the last 10 years, while less and less time is allotted to the teaching of evolution and systematics on the undergraduate level (Clarke 1990).

For those who work on certain groups of organisms, a few academic and applied systematics jobs still exist (the classified advertisement section of *Science* yields a few ads per month for systematic botanists, for example). But for the field I know best, invertebrate zoology, prospects for a student interested in a career in systematics are poor, indeed. Little invertebrate zoology is taught in college and university biology departments; most are molecular or medically oriented. The field may still be entered in the few universities with a marine biology major or via summer courses given at marine laboratories. Those college or university professors who utilize marine invertebrates in their research tend to be either ecologists or developmental biologists, for ecology and embryology are still part of most biology curricula (Heppner et al. 1990). They naturally tend to draw their students toward research in those disciplines. They are also reluctant to encourage a student toward systematics because of the limited number of job openings.

Jobs in invertebrate systematics are pretty much limited to research museums, which are enduring hard times. A number of museums have recently contracted their staff or closed entirely. For example, recent cuts at the British Museum canceled research on seven major marine groups: echinoderms, sponges, coelenterates, bryozoans, and annelids (Anonymous 1990). The San Diego Museum recently let 40% of its staff go to resolve financial difficulties (Marshall 1991). Most museums can hire curators only to replace losses through death or retirement. There is an understandable tendency to hire someone specializing in a group for which the museum already has a considerable collection. This makes specializing on a poorly known group (even though the research potential might be great) an even greater risk. Under such conditions perhaps it is surprising, not that there are so few specialists, but that there are *any* specialists left at all.

Systematics and Marine Conservation

The introductory examples in this paper show how serious the situation is at present. However, in contrast to the sense of crisis over tropical habitats, there is a belief among marine biologists and conservationists that there is still time for documentation of marine diversity and planning for its conservation (Carlton 1989; Vermeij 1989; Thorne-Miller and Catena 1991).

Documentation of the Basic Resource

The next section of this paper pointed out that there are a number of ways to measure marine diversity and that the proportion of earth's diversity that is marine in habitat is high by all of them. Some authors (e.g., Ray 1988) have argued that functional or ecological diversity is the most significant criterion and that marine management plans can be worked out on that basis, without spending time and money in attempting to inventory and describe the still unknown members of marine ecosystems.

This aspect of diversity is certainly significant and should be addressed in all conservation plans, but as a systematist I must argue that the taxonomic aspect—or rather our massive ignorance of the taxonomic aspect—of the existence and identity of many of the components of marine ecosystems is too important to ignore. Only for a few groups, in the shallowest and best studied parts of the sea, can we say that we know 80% of the species that exist. For many groups, in most of the oceans, we have inventoried less than 50%. For all but the largest and most commercially valuable, we know almost nothing of their ecology, their life history patterns, or their functional role. Any realistic management plan requires a better database than this.

Documentation of Threatened Species

We know that we are having a negative impact on the seas and that extinctions are likely to increase. First we need to know what species are at risk. Closest to extinction are the large marine mammals (like the Mediterranean Monk Seal) and those shorebirds whose wetland habitat is disappearing, but as marine habitats become increasingly disrupted, smaller organisms will be affected, too. Editors of the *Red Data Book* compilation created some new categories for invertebrates—categories that reflect our ignorance even of crucial aspects of their life histories: *vulnerable* (declining populations, likely to move into the endangered category), *commercially threatened, threatened community* (groups of ecologically linked taxa in a defined area like a coral reef), *threatened phenomenon* (e.g., migrating species), and *indeterminate* (those we just do not have enough information on to categorize) (Collins and Wells 1983).

But of course, if you never knew a species was there in the first place, you would never notice it had become extinct. The phenomenon of "taxonomy as destiny" was recently illustrated for a terrestrial reptile—the new Zealand tuatara *Spenodon* (Daugherty et al. 1990; May 1990). Two species of tuatara, one of them with several subspecies, were recognized in 1877, but legislation (starting in 1895) recognized only one species. As a result present populations of the second species are very small and limited to one island, on which they appear threatened, and one subspecies of the common species has become extinct. In view of our limited knowledge of marine species, it is more likely that similar events are occurring in the sea.

Documentation of Marine Habitats

Quantitative systematic and ecological inventories providing detailed knowledge of the plants and animals present in a habitat are critical for any kind of management. Most of us assume these baseline data were collected in the nineteenth or early twentieth century, the great age of floristic and faunistic studies, but in most of the oceans, even in most of the shallower portions of the oceans, these were not collected. Even where the surveys did take place their results may not be useful today (because, for example, the necessary concurrent measurement of physical and chemical parameters did not take place). The famous expeditions of the nineteenth century, like the Dutch *Siboga* Expedition or the British *Challenger* Expedition, did much to form a basis for the sciences of oceanography and marine biology, but their sampling was not quantitative and often lacked precise locality data (Soule 1988).

Reliance on old reports can be unwise for another reason, also. Usually the early workers had only a limited selection of literature available and often identified tropical or new world forms with European species, when, in fact, they were new species. While the old reports can be useful for those engaged in present-day ecological and systematic studies, they cannot by themselves provide the kind of database we need today.

What is more, even when such survey work was being done, it was seldom thought desirable to survey home waters—the more exotic, the farther from home, the deeper, the better the chance of receiving funding—even then. As a result, many of the most urbanized coastal areas, the very areas in which human impact has been greatest, have never been surveyed. The practice continues today—if systematics has any glamour at all, it is only when carried out in the most exotic locale. In my own experience, I found it far easier to get support to study the systematics of Antarctic bryozoans than those of the northeastern U.S. coast—even though the local fauna was less well known.

Recognition of Introductions

Even marine biologists may not recognize man's impact on marine environments, even on shorelines we think of as unaltered, like those of marine and estuarine reserves. Numerous marine communities have been greatly changed by the introduction of exotic species, many of which took place before the biological surveys of the nineteenth century (Carlton 1989). Some of those introductions were purposeful. For example, the rocky coast of New England is structured as we know it, free of algae and the sediment deposition that would promote marsh development, by the activity of one introduced species, *Littorina littorea,* the common periwinkle, which was brought to Nova Scotia about 1840 as a food source (Fleming 1990).

Other introductions were accidental, via organisms fouling ships' hulls, attached to ballast material, or sucked up into ballast water as larvae, to be released hundreds or thousand of kilometers from their origin. For example, the ballast water of ships arriving in an Oregon harbor from Japan was found to contain more than 200 species of living zooplankton and phytoplankton. Often these immigrants were present in large enough numbers to establish themselves permanently in the new area (Carlton 1985; Williams, Griffiths, Van der Wal, and Kelly, 1988).

Exotic marine species may have a profound effect on the communities in which they arrive, not so much by catastrophic extinctions of native species (as often occur when exotic species are introduced in freshwater or terrestrial environments), as by changes in ecological interactions and species abundances. In San Francisco Bay, for example, the Asian clam *Potamocorbula,* introduced since 1985, has been spreading rapidly. A recent study showed that it may have already permanently changed benthic community dynamics in at least one region of the Bay (Nichols, Thompson, and Schemel 1990).

The realization that many present-day distributions are not the results of dispersal or vicariance events but instead have been (geologically speaking) instantaneous came, not from ecological studies of a single community, but from the efforts of systematists working on diverse groups of invertebrates, whose knowledge of distribution patterns and species histories led them to perceive the cosmopolitan aspects of the shallow water marine biota (Carlton 1989).

The Future

Conserving marine biodiversity and managing marine resources intelligently will take cooperative efforts by systematists, ecologists, and applied biologists, as well as by concerned citizens.

Awareness

The first need is to promote public awareness of the problem. Magazines and newspapers do report the mass deaths of large appealing marine mammals like dolphins and seals, e.g., *Time,* Aug. 1, 1988, cover story, "The Dirty Seas," (Toufexis 1988), or the fate of sea turtles mistaking balloons for jellyfish, e.g., *New Jersey Audubon,* "Death by Plastic," (Cimbal 1990), but even many marine biologists do not recognize the extent to which the oceans have been altered by human activity. Most marine biology and ecology texts still contain nothing on marine introductions; some do not include even a chapter on resource conservation (Carlton 1989). It is no wonder that professional biologists, students, and the general public do not realize that there is still a great need just to find out what lives in the oceans.

Systematists can play an important role in alerting the world to the problems faced by marine species. For example, the last representative of the extinct coelocanth fishes, *Latimeria,* was discovered in 1938. Its only known habitat is the deep submarine slope off the Comores Archipelago near Madagascar. As a British systematist recently pointed out, the use of more sophisticated fishing methods in the last few years dramatically increased the number of coelocanths caught, yet almost nothing was known of its life history, diet, or habitat requirements, and thus there was no way to determine whether the increased number of catches was pushing this living fossil to extinction (Forey 1988).

In this case, a group was formed to oversee the fish's conservation, so the species may survive. But not only rare fish like the coelocanth are threatened. Overexploitation is the greatest problem facing commercial fisheries today. Development and deployment of high-technology equipment, like miles-long plastic driftnets, with satellite-transmitted location signals, has far outpaced fishery knowledge and controls (see McCloskey 1991).

The highly efficient fishing methods used today kill more fishes in heavily exploited populations (like those of cod or haddock) than natural causes do. Systematists and evolutionary biologists can show how different exploitation regimens can actually affect the way populations evolve and may be able to point out how selection can be controlled to increase, rather than decrease, the yield (Law 1991).

Systematic studies at the population level can be essential to manage and preserve threatened species. For example, study of mitochondrial deoxyribonucleic acid (DNA) patterns in green turtles showed little gene flow between populations from different home beaches (the birth beaches to which females return year after year to lay their eggs), which means that conserving one population does not help any of the others. Conservation efforts must be handled separately for each population (Meylan, Bowen, and Avise 1990).

Systematists must become active also in educating the public about the role museums play. Most people, even well-educated people, are unaware that museums are not just extra-large cabinets of specimens, preserved for their beauty or curiosity. Museums are scientific research centers and their specimens are an important database, not just for service identifications, but for all kinds of systematic research and to document environmental change and human impact (Hoagland 1990).

Even other scientists are often ignorant about what systematists do. They do not realize that museum-based systematists, although they have varying amounts of responsibility to collections management, curation, and education, are evaluated, just like other academic scientists, on the quality and quantity of the research they publish in peer-reviewed journals. What outsiders often see as a curator's chief function—the identification of specimens for others—takes a very low priority and a small percentage of their time (Winston 1988).

As the second section of this paper showed, probably half of the diversity of ocean life still remains to be cataloged. Whereas the dwindling number of taxonomists who study marine organisms was mentioned, and the need for expanding their ranks by training new taxonomists was implied, some aspects of research policy and job structure may have to be changed to make it possible for those now working in museums to take on these tasks effectively. Many groups of invertebrates and algae still urgently need the basic kinds of taxonomic work: checklists, inventories, area surveys, floral and faunal studies, etc. Museum systematists receive salaries but usually little or no field money from institutional sources. They cannot pursue such field studies without grant support. Yet this basic aspect of systematics is considered the least creative by both granting agencies and promotion committees. This creates a dilemma for a person specializing in such a group—knowledge and conscience may say this work is necessary—but common sense says it is career suicide.

A way must be found to reward specialists for pursuing practical and basic work as part of their career, before these scientists can make their strongest contributions both to pursuing such studies themselves and to training others to pursue them.

Inventory

We know that diversity has a real survival value for ecosystems by proving stability and buffering communities against extinction. We have learned that the hard way, again and again, as we have destroyed one ecosystem after another. This survival value outweighs any economic value we can come up with (Ehrenfeld 1986).

Various attempts have been made to find methods to avoid having to tediously inventory this diversity in order to understand and protect the system. Bioindicators, biomarkers, and keystone species have all been promoted as solutions. These methods can be extremely useful for biological monitoring, but they are all based on assumptions about the system that may or may not hold, and they must be used cautiously, in well-defined systems to give accurate results (Soule 1988; McCarthy and Shugart 1990).

Ultimately, there is no alternative to inventory for understanding the diversity of marine ecosystems—we just have to get in there and do it. This does not mean that we have to wait until the final details of description and classification are worked out, but we *have* to get at least to the 80%, rather than the 50% level, for the whole ocean. This should not be impossible.

Recently four tropical biologists have claimed that they can go into a tropical area, decide what parts of it are most worth preserving, and identify many of the species present within a few days. The highly trained ornithologist can identify birds by their songs, without needing to see them, while the botanist can recognize tree species from the air, etc. (Roberts 1991). Most marine animals do not cooperate by producing identifying songs, but this "taxo-quick" approach could work in marine habitats, too.

Such an approach would not take the place of an exhaustive inventory. It *could* identify the most important areas to preserve for the future. Teams would have to be larger, to deal with the much larger number of groups, and stay longer (as daily underwater time is limited), but the team approach could be very effective. An expert on a group, even if unable to name all the species in a locality (because not all have been named), can nevertheless quickly (1) provide a reckoning of approximately how many are there, (2) point out which ones are probably unique or endemic, (3) identify the species that have been described, and (4) provide preliminary information on the ecological role of the most abundant species.

While important taxonomic discoveries can still be made for many groups of marine organisms using the most primitive equipment (e.g., a shovel at low tide), major survey work needs access to research vessels and/or marine laboratories with small boat and diving facilities. For deep-water and pelagic organisms major discoveries will depend on access to latest technology for underwater exploration: research submarines or ROVs equipped with video systems and collecting arms. Yet the cost of such operations is so minimal, compared with that of a big medical or molecular project, for example, that it is obvious that what stops us is not economics but attitude.

New programs for cataloging tropical diversity (like Costa Rica's INBio Program) are training "parataxonomists" to help with that enormous job. These parataxonomists are people, not necessarily with previous scientific training, who receive an intensive six-month course that enables them to collect, prepare, and label specimens (Tangley 1990). The success of this program (INBio's first class added more insect specimens to the National Museum than had been deposited there in the past hundred years), as well as my own experience in training American Museum of Natural History volunteers (many of them likewise without previous scientific background) to sort, label, and even give preliminary identifications to marine bryozoans, makes me believe that there is great potential in this approach for marine conservation. Such an approach has the added benefit of creating a large network of committed people at the grass-roots level, both to aid in management and to agitate for better legislation.

Management

There is a reasonable amount of literature on the conservation of marine ecosystems, though not nearly as much as for terrestrial and freshwater systems (see reviews by Hatcher, Johannes, and Robertson 1989; Thorne-Miller and Catena 1991). But the very nature of the fluid medium of seawater means that the problems of marine ecosystems cannot be solved just by applying techniques effective on land. Many marine organisms spawn freely into seawater and produce larval or juvenile stages that spend days or weeks as drifting plankton. It is useless to provide a reserve to protect adults, for example, if population recruitment takes place from parental stock miles distant and that parental population is not protected, too. Distant events can also have rapid and drastic effects on pristine or protected populations when their effects are carried by water—as recent disastrous oil spills have shown us (Davidson 1990; Thorne-Miller and Catena 1991).

Protection of habitats works best when those habitats are geographically limited and self-sustaining, as with some coral reef areas. This is the opposite situation from that found in most coastal habitats, which are filled with migrating birds, fish, and crustaceans, as well as stocked by drifting plankton and larvae.

Clearly, management of marine ecosystems must be based on an integrated system of activities and controls: we must (1) find out what is there, (2) create a network of protected areas and reserves, and (3) ameliorate pollution of the connecting water, or all our work is wasted.

Workable solutions have to come via legislation and regulation because— as much we like to talk about saving our habitat—most of us do only what we are required to do by law.

Nor can effective solutions be limited by local or national boundaries; they depend on cooperation across boundaries to deal with the biological realities of life in seawater. These cooperative policies must include regulation and abatement of marine pollution, whether its ultimate source is the land or the sea; regulation and management of marine resources in a sustainable manner; control of introductions of exotic marine organisms; establishment of protected areas; and the implementation of economic incentives and sanctions (Thorne-Miller and Catena 1991).

Legislation

While we are still far from that point, at least there is some evidence that people and governments are beginning to change their thinking. U.S. regulations dealing with marine life have been fragmented and confusing. Marine mammals, migratory seabirds, and threatened and endangered species have come primarily under the jurisdiction of the federal government, while the states have had responsibility for all "nongame wildlife," a category including most other animals and (sometimes) plants. Laws and programs are scattered and confusing, and the agencies that oversee them may have no control over most of the habitats in which the organisms live. At last the situation seems to be changing. More states are attempting to give up the old, illogical definitions, in favor of more nearly comprehensive systems for the protection of all wildlife, although terrestrial habitats and organisms still seem to be favored over marine ones (Vickerman 1990). Just the fact that a bill to create a Department of the Environment at cabinet level could reach Congress (Hair 1990) indicates there is hope of real change.

International organizations have been formed to deal with some of these problems. For example, we now have the LDC (Convention on the Prevention of Marine Pollution by the Dumping of Wastes and Other Matter, London, 1972), MARPOL (Convention on the Prevention of Marine Pollution by Dumping from Ships and Aircraft, Oslo, 1972) (for decreasing pollution from ships), United Nations Convention on the Law of the Sea, 1982 (which attempts to set up a system of international marine law but which so far has been ratified by only 15 of the 60 nations necessary), the International Whaling Commission, UNEP (the United Nations Environmental Program, including the Regional Seas Program), the IUCN (International Union for the Conservation for Nature and Natural Resources), and CCAMLR (the Convention for the Conservation of Antarctic Living Resources). These organizations have definitely had some effect, but they need strengthening, particularly by (1) greater consensus about what protection of a resource or an area means, (2) specific addressing of biological diversity and its protection, and (3) more effective means of enforcement (Cognetti 1990; Thorne-Miller and Catena 1991).

Advocacy

Advocacy is important, too. Nongovernmental organizations like national and international environmental groups can be very effective both in seeing that regulations are enforced and in protecting and restoring marine habitats. We have seen again and again in the terrestrial environment how citizen involvement at the community level can bring about major changes in policy. All those dependent on the sea, from school children, to fishermen, to coastal residents to tourists, to marine biologists, have the power to take action to influence the policy makers.

Systematics is essential to marine conservation. However, systematics by itself cannot solve the problem—any solution requires cooperation with other disciplines in biology and even outside science entirely. Systematics is essential, but the results of systematic studies must be integrated into the whole process of ecological study, monitoring, and management, if the diversity of the oceans is to be preserved. This cannot be done without changes in attitude, laws, and policy. I urge everyone who reads this to get involved in marine conservation, either as scientist or citizen or both—like it or not—the planet is two thirds ocean. If we destroy our dry third, the global cycles that control our biosphere may persist, but if we destroy the ocean, that is the end for life on earth.

REFERENCES

Anonymous. 1990. Fate of the Natural History Museum, London. *Association of Systematic Collections Newsletter* 18:38.

Baden, S. P., L. -O. Loo, L. Pihl, and R. Rosenberg. 1990. Effects of eutrophication on benthic communities including fish: Swedish west coast. *Ambio* 19:113–122.

Barnes, R. D. 1987. *Invertebrate Zoology,* 5th ed. Philadelphia: Saunders.

Barnes, R. D. 1989. Diversity of organisms: How much do we know? *American Zoologist* 29:1075–1084.

Barnes, R. S. K. 1984. *A Synoptic Classification of Living Organisms.* Sunderland, Mass.: Sinauer Associates.

Barnes, R. S. K. and R. N. Hughes. 1982. *An Introduction to Marine Ecology.* London: Blackwell.

Beaumont, A. R., P. B. Newman, D. K. Mills, M. J. Waldock, D. Miller, and M. E. Waite. 1989. Sand-substrate microcosm studies on tributyl tin (TBT) toxicity to marine organisms. *Scientia Marina* 53:737–743.

Beder, S. 1990. Sun, surf and sewage. *New Scientist* (July 14) 1990:40–45.

Bennett, D. W. 1990. Fish. *Underwater Naturalist* 19(3):32.

Bergquist, J., W. Giebel, and D. Rudman. 1990. Tuna and plastics. *Underwater Naturalist* 19:18–19.

Brown, B. E., ed. 1990. Coral Bleaching. *Coral Reefs,* 8(4):153–232.

Carlton, J. T. 1985. Transoceanic and interoceanic dispersal of coastal marine organisms: The biology of ballast water. *Oceanographic Marine Biology Annual Review* 23:313–371.

Carlton, J. T. 1989. Man's role in changing the face of the ocean: biological innovations and implications for conservation of near-shore environments. *Biological Conservation* 3:265–273.

Cherfas, J. 1990. The fringe of the ocean—under siege from the land. *Science* 248:163–165.

Cimbal, K. 1990. Death by plastic. *New Jersey Audubon* 16:13–14.

Clarke, B. 1990. Systematics and evolutionary biology in Britain. *Association of Systematic Collections Newsletter* 19(4):58–59.

Cognetti, G. 1990. Conservation of marine environments in the Mediterranean. *Marine Pollution Bulletin* 21:115–117.

Cohen, D. M. 1970. How many recent fishes are there? *Proceedings of the California Academy of Sciences, 4th series* 38:341–346.

Collins, M. and S. Wells. 1983. Invertebrates—who needs them? *New Scientist* 98(1358):441–444.

Daugherty, C. H., A. Cree, J. M. Hay, and M. B. Thompson. 1990. Neglected taxonomy and continuing extinctions of tuatara (*Sphenodon*). *Nature* 347:177–179.

Davidson, A. 1990. *In the Wake of the Exxon Valdez*. San Francisco: Sierra Club Books.

Dietz, R., M. -P. Heide-Jorgensen, and T. Harkonen. 1989. Mass deaths of harbor seals (*Phoca vitulina*) in Europe. *Ambio* 18:258–264.

Drew, L. 1990. Truth and consequences along oiled shores. *National Wildlife* 28: 34–42.

Edwards, S. R., G. M. Davis, and L. I. Nevling. 1985. The systematics community. Lawrence, Kansas, Association of Systematics Collections.

Ehrenfeld, D. 1986. Thirty million cheers for diversity. *New Scientist* 110 (1512): 38–43.

Erwin, T. L. 1983. Beetles and other insects of tropical forest canopies at Manaus, Brazil, sampled by insecticidal fogging. In S. L. Sutton, T. C. Whitmore, and A. C. Chadwick, eds., *Tropical Rain Forest: Ecology and Management*, pp. 59–75. Oxford: Blackwell.

Fang, C. S. 1990. Petroleum drilling and production operations in the Gulf of Mexico. *Estuaries* 13:89–97.

Fenaux, R. and M. J. Youngbluth. 1990. A new mesopelagic appendicularian, *Mesochordaeus bahamasi* gen. nov., sp. nov. *Journal of the Marine Biological Association of the United Kingdom* 70:755–760.

Fenchel, T. 1988. Marine plankton food chains. *Annual Review of Ecology and Systematics* 19:19–38.

Fleming, C. B. 1990. A snail's landscape. *Sea Frontiers* 36(6):51–55.

Fletcher, F. L., M. A. Shears, M. J. King, P. L. Davies, and C. L. Hew. 1988. Evidence for antifreeze protein gene transfer in Atlantic salmon (*Salmo salar*). *Canadian Journal of Fisheries and Aquatic Science* 45:352–357.

Forey, P. 1988. Golden jubilee for the coelocanth *Latimeria chalumnae*. *Nature* 336:727–732.

Gibbs, P. E., P. L. Pascoe, and G. R. Burt. 1988. Sex change in the female dog-welk, *Nucella lapillus*, induced by tributyltin from anti-fouing paints. *Journal of the Marine Biological Association of the United Kingdom* 68:715–732.

Grzimek, B., ed. 1968. *Grzimek's Animal Life Encyclopedia*, vol. 6. New York: Van Nostrand Reinhold.

Guzman, H. L. M., J. Cortes, P. W. Glynn, and R. H. Richmond. 1990. Coral mortality associated with dinoflagellate blooms in the eastern Pacific (Costa Rica and Panama). *Marine Ecology Progress Series* 60:299–303.

Hair, J. D. 1990. Elevating EPA to cabinet rank. *Environmental Science Technology* 24:1143.

Hansson, S. and L. G. Rudstam. 1990. Eutrophication and Baltic fish communities. *Ambio* 19:123–125.

Harwood, J. 1989. Lessons from the seal epidemic. *New Scientist* 121 (1652):38–42.

Hatcher, B. G., R. E. Johannes, and A. I. Robertson. 1989. Review of research relevant to the conservation of shallow tropical marine ecosystems. *Oceanographic Marine Biology Annual Review* 27:337–414.

Heppner, F., C. Hammen, G. Kass-Simon, and W. Krueger. 1990. A *de facto* standarized curriculum for US college biology and zoology. *BioScience* 40:130–134.

Heydorn, A. E. D. 1989. The conservation status of Southern African Estuaries. In B. J. Huntley, ed., *Biotic Diversity in Southern Africa,* pp. 290–327. Cape Town: Oxford University Press.

Hoagland, K. E. 1990. Natural history museums and the public trust. *Environment West* 1:4–5.

Holligan, P., ed. 1990. *Coastal Ocean Fluxes and Resources.* Stockholm: International Geosphere-Biosphere Programme Global Change Report no. 14, pp. 1–53.

Irwin, H. S., W. W. Payne, D. M. Bates and P. S. Humphrey. 1973. America's systematic collections: a national plan. Lawrence, Kansas, Association of Systematics Collections.

Isleib, M. E. 1989. Of dead zones and the sound of silence. *American Birds* 43: 215–217.

Joyce, C. 1989. Poisonous algae killed the Atlantic dolphins. *New Scientist* (February 11) p. 31.

Kennicut, M. C. et al. 1990. Oil spillage in Antarctica. *Environmental Science and Technology* 24:620–624.

Kristensen, R. M. 1983. Loricifera: A new phylum with Aschelminthes characters from the meiobenthos. *Zeitschrift für Zoologische Systematik und Evolutionsforschung* 21:163–180.

Lane, N. G. 1990. A census of past and present life. *Journal of Geological Education* 38:119–122.

Law, R. 1991. Fishing in evolutionary waters. *New Scientist* (March 2), pp. 35–37.

Lean, G., D. Hinrichsen, and A. Markham. 1990. *Atlas of the Environment.* New York: Prentice Hall.

Leblond, P. H. 1990. The role of cryptozoology in achieving an exhaustive inventory of the marine fauna. *La mer* 28:1–4.

Lenihan, H. S., J. S. Oliver, and M. A. Stephenson. 1990. Changes in hard bottom communities related to boat mooring and tributyltin in San Diego Bay: a natural experiment. *Marine Ecology Progress Series* 60:147–159.

Marshall, E. 1991. Hard times for San Diego Museum. *Science* 251:375.

May, R. M. 1988. How many species are there on earth? *Science* 241:1441–1449.

May, R. M. 1990. Taxonomy as destiny. *Nature* 347:129–130.

McCarthy, J. F. and L. R. Shugart, eds. 1990. *Biomarkers of Environmental Contamination.* Chelsea, Michigan, Lewis Publishers.

McCloskey, W. 1991. Casting drift nets with the squidders. *International Wildlife* 21(2):40–46.

Meylan, A. B., B. W. Bowen, and J. C. Avise. 1990. A genetic test of the natal homing versus social facilitation models for green turtle migration. *Science* 248: 724–727.

Michel, J. 1990. The *Exxon Valdez* oil spill: Status of the shoreline. *Geotimes* 35(5):20–22.

Miller, J. E. and D. L. Pawson. 1989. *Hansenothuria benti,* new genus, new species (Echinodermata: Holothouroidea) from the tropical western Atlantic: A bathyal, epibenthic holothurian with swimming abilities. *Proceedings of the Biological Society of Washington* 102:977–986.

Milne, R. 1990. Dutch call for fishing-free zone in the North Sea. *New Scientist* 125(1708):26.

Moss, S. T., ed. 1986. *The Biology of Marine Fungi.* Cambridge, U.K.: Cambridge University Press.

Nichols, F. H., J. K. Thompson, and L. E. Schemel. 1990. Remarkable invasion of San Francisco Bay (California, USA) by the Asian clam *Potamocorbula amurensis.* II. Displacement of a former community. *Marine Ecology Progress Series* 66:95–101.

Odum, E. P. 1989. *Ecology and Our Endangered Life-Support Systems.* Sunderland, Mass.: Sinauer Associates.

Pain, S. 1990. Coral reefs will thrive in the greenhouse. *New Scientist* 125(1706):30.

Parker, S. P., ed. 1982. *Synopsis and Classification of Living Organisms,* 2 vols. New York: McGraw-Hill.

Pedersén, M., M. Björk, C. Larsson, and S. Söderlund. 1990. Ett marint ekosystem i obalans—dramatsika förändrinagar av hårdbottnarnans växtsamhällen. *Fauna och Flora* 85:202–211.

Podolsky, R. H. 1989. Entrapment of sea-deposited plastic on the shore of a Gulf of Maine Island. *Marine Environmental Research* 27:67–72.

Priestley, A. J. 1990. Sydney and oceanic sewage disposal. *Search* 21:243–248.

Raloff, J. and R. Monastersky. 1991. Gulf oil threatens ecology, maybe climate. *Science News* 139(5):71–73.

Ray, G. C. 1988. Ecological diversity in coastal zones and oceans. In E. O. Wilson, ed., *Biodiversity,* pp. 36–50. Washington, D.C.: National Academy Press.

Roberts, L. 1990. Warm waters, bleached corals. *Science* 250:213.

Roberts, L. 1991. Ranking the rainforests. *Science* 251:1559–1560.

Rosenberg, R., O. Lindahl, and H. Blanck. 1988. Silent spring in the sea. *Ambio* 17:289–290.

Sathyendranath, S., A. D. Gouveia, S. R. Shetye, P. Ravindran, and T. Platt. 1991. Biological control of surface temperature in the Arabian Sea. *Nature* 349(6304):54–56.

Scudder, G. G. E. 1987. The next 25 years: Invertebrate systematics. *Canadian Journal of Zoology* 65:786–793.

Sheppard, C. and A. Price. 1991. Will marine life survive the Gulf War? *New Scientist* 129(1759):36–40.

Simons, M. 1990. Virus linked to pollution is killing hundreds of dolphins in Mediterranean. *The New York Times* (October 28) p. L5.

Soule, D. F. 1988. Marine organisms as indicators: Reality or wishful thinking. In D. F. Soule and G. S. Kleppel, eds., *Marine Organisms as Indicators*, pp. 1–11. New York: Springer-Verlag.

Spence, S. K., S. J. Hawkins, and R. S. Santos. 1990. The mollusc *Thais haemostoma*—an exhibitor of 'imposex' and potential biological indicator of tributyltin pollution. Publicazioni della Stazione di Nopoli. I. *Marine Ecology* 11:147–156.

Stace, C. 1989. *Plant Taxonomy and Biosystematics*, 2d ed. London: Edward Arnold.

Steere, W. C. et al. 1971. *The Systematic Biology Collections of the United States: An Essential Resource. Part I: The Great Collections—Their Importance, Condition, and Future.* The Bronx, N.Y.: New York Botanical Garden.

Stuessy, T. F. and K. S. Thompson, eds. 1981. *Trends, Priorities, and Needs in Systematic Biology.* Lawrence, Kansas: Association of Systematic Collections.

Tangley, L. 1990. Cataloging Costa Rica's diversity. *BioScience* 40:633–636.

Thorne-Miller, B. and J. Catena 1991. *The Living Ocean: Understanding and Protecting Marine Biodiversity.* Washington, D.C.: Island Press.

Toufexis, A. The dirty seas. *Time* 132(5):44–50.

Turner, M. H. 1990. Oil spill: Legal strategies block ecology communications. *Bioscience* 40:238–242.

Underdal, B., O. M. Skulberg, E. Dahl, and T. Aune. 1989. Disastrous bloom of *Chyrsochromulina polylepis* (Prymnesiophyceae) in Norwegian coastal waters 1988—mortality in marine biota. *Ambio* 18:265–270.

Vermeij, G. J. 1989. Saving the sea: What we know and what we need to know. *Conservation Biology* 3:240–241.

Vickerman, S. E. 1990. State wildlife protection efforts. In *Preserving Communities and Corridors*, pp. 67–96. Washington, D.C.: Defenders of Wildlife.

Williams, R. J., F. B. Griffiths, E. J. Van der Wal, and J. Kelly. 1988. Cargo vessel ballast water as a vector for the transport of non-indigenous marine species. *Estuarine and Coastal Shelf Science* 26:409–420.

Williamson, P. and P. M. Holligan. 1990. Ocean productivity and climate change. *Trends in Ecology and Evolution* 5:299–303.

Wilson, E. C. 1990. Mass mortality of the reef coral *Pocillopora* on the south coast of Baja California Sur, Mexico. *Bulletin of the Southern California Academy of Sciences* 89:39–41.

Wilson, E. O. 1985. Time to revive systematics. *Science* 230:1227.

Wilson, E. O. 1988. The current state of biological diversity. In E. O. Wilson, ed., *Biodiversity*, pp. 3–18. Washington, D.C.: National Academy Press.

Wilson, E. O. 1991. Funding unsexy science. *Science* 251.

Wilson, T. R. S. 1989. Climate change: Possible influence of ocean upwelling and nutrient concentration. *Terra Nova* 1(2):172–176.

Winston, J. E. 1988. The systematists' perspective. *Memoirs of the California Academy of Sciences*, no. 13:1–6.

Woolley, T. A. 1986. *Acarology; Mites and Human Welfare.* New York: Wiley.

Youngbluth, M. J., T. G. Bailey, and C. A. Jacoby. 1990. Biological explorations in the mid-ocean realm: Food webs, particle flux, and technological achievements. In Y. C. Lin and K. K. Shida, eds., *Man in the Sea*, 2:191–208. San Pedro, Calif.: Best Publishing.

11 : The Conservation of Animal Diversity in Cuba

Gilberto Silva Taboada

Phenomena such as the greenhouse effect, the depletion of the ozone layer, or the acidification of lakes and forests obviously relate to the conservation of biological diversity on a global scale; therefore, effective action at this level demands a great deal of international cooperation.

At the same time, on a local scale there are other sorts of environmental problems affecting biological diversity that any country must face individually. I intend to offer a brief overview of some of these local problems as regards the conservation of native terrestrial animals in Cuba, the country I come from and the area of conservation with which I am most familiar.

What makes Cuba a faunal conservation issue of any particular interest to this symposium?

According to recent listings, roughly 15,000 species, comprising more than 10,000 arthropods, 4,000 other invertebrates, and 500 vertebrates, are known from Cuba, and this round figure is estimated to represent not more than 60% of the actual native fauna.

Since Cuba is a truly oceanic archipelago, rates of speciation have been greater than in other territories of comparable size and ecological diversity. Hence, evolutionary radiation favored the proliferation of endemic species whereby faunal groups such as amphibians, reptiles, mollusks, and the majority of arachnids, millipedes, and insects have attained levels of endemism ranging from 80% to 95%.

It is a well-established fact that insular faunas are highly vulnerable to environmental stress and that in such circumstances endemic species are among the first to disappear. Therefore, it is not by accident that endemic species contribute at present up to 88% of the 79 taxa of Cuban land vertebrates known as extinct or extirpated forms, as well as up to 85% of the 111 extant taxa currently subject to varying degrees of extinction risk throughout the territory. Aside from these statistics, little is known of the conservation status of the highly endemic invertebrate fauna.

On the other hand, Cuba belongs to the group of nations euphemistically called less developed countries, where poverty and ignorance severely exacerbate environmental deterioration, in ways that more developed countries do not usually experience. Overcoming poverty and ignorance is certainly a sine qua non condition for ecological welfare, though not a guarantee in itself that such a desirable state will be accomplished.

Therefore, I have chosen to introduce in my analysis the well-known fact that Cuba has struggled against poverty and ignorance for the past 30 years in a way unprecedented among Third World countries.

The experience of three decades of socialized economy in Cuba illustrates a way of coping with environmental problems that is unique in Latin America. Within this scope, I focus on the socioeconomic and political contexts in which these problems—in particular, the conservation of animal diversity—are presently approached in Cuba.

The Socioeconomic Context

As is well known, habitat modification is the main cause of decline in animal diversity worldwide.

In 1959, when a popular revolution came to power in Cuba, the forest cover had been reduced to 14% of the island's territory and a wide range of environmental calamities had already occurred as a result of irrational exploitation of natural resources by previous regimes.

Nevertheless, after 1959, the rate of environmental modification sped up because of two fundamental reasons: (1) under the new government, far-reaching economic and social transformations took place in a very short time; (2) regardless of the rate of socioeconomic development, human capacity to impair the environment reaches progressively farther into time and space, and these dimensions defy prediction. Thus, in its latest edition (1989), the *National Atlas of Cuba* includes a classification of human-caused modification of landscapes wherein the category identified as "highly modified landscape" has almost tripled its area since 1959.

Some comparative figures may clarify the magnitude of economic and social transformations in Cuba over the past 30 years, particularly those directly or indirectly bearing upon the quality of the natural environment and the diversity of the autochthonous fauna dependent on it.

With an available territory of almost 111,000 square kilometers, the population of Cuba has increased to such an extent that there are now 92 inhabitants per square kilometer, a ratio comparable to that of countries with the highest demographic density. This has, of course, been greatly influenced by the fact that medical care in Cuba has become what international health organizations currently acknowledge as the best public health service in Latin America. The rate of neonatal mortality has been reduced from 60 per every 1,000 live births in 1959 to 13 in 1989. During that same period, life expectancy went from 55 to 74 years. Consequently, demographic density is becoming a major factor in environmental pressure.

The urge to meet undeferrable needs of the population, which increased along with those of foreign trade, led the new administration to implement extensive agricultural, industrial, and urban development plans, with the predictable outcome that environmental damage also accelerated. Thus air and water pollution substantially increased, as did soil erosion and degradation of remnant forests, among other factors of ecosystem destabilization.

Thousands of new human settlements were set up, and highways and roads, which currently add up to 30 times the length of the island of Cuba, were constructed. An extensive damming program increased the reservoir potential from 48 million cubic meters in 1959, to more than 6 billion in 1989.

Presently, 70% of the national territory is devoted to agriculture and cattle-raising, and natural and seminatural areas remain only in some mountainous or coastal regions. The reforestation effort has made it possible to increase forest cover up to 18% of the territory, but population growth has lowered the relative ratio to 0.9 hectare of forest area per inhabitant, one of the lowest in Latin America.

However, the fact that early in the 1960s thousands of new jobs were created had a positive immediate effect on one of the components of the acute degradative process that forests were undergoing at that time. Those segments of the rural population living under extremely marginal conditions benefited from better living conditions. Harmful practices related to subsistence agriculture and the production of charcoal—at that time the main fuel for Cuban homes—were modified or eliminated. Within a couple of years, charcoal cooking stoves had been replaced nationwide by industrial household appliances.

Education did not lag behind the thrust given to the economy and public health. As a first step, in only one year, 100,000 young people taught elementary reading and writing to half a million adults among the rural population, 42% of which was illiterate in 1959. In the following years, education underwent a colossal boost at the primary, secondary, and technological levels, and higher learning was diversified and reoriented to suit the country's prospective development. (In Cuba in 1959, there were more lawyers than biologists, veterinarians, chemists, physicists, and engineers of every kind, all taken together.) From 1959 to 1989, the overall number of Cuban students rose from 800,000 to more than 3 million.

With the establishment of the Academy of Sciences, in 1962, the foundation was laid for developing the scientific and technical potential called for by new economic and social needs.

Electric power output was substantially increased, and it brought radio, television, and several other educational, cultural, and recreational services to the remotest corners of the country.

As time went by, an absolute majority of the population had acquired enough purchasing power and spare time to be able to afford tourism. Then, the construction of hotels and other kinds of recreational facilities was stepped up in numerous spots throughout the Cuban countryside. In 1981, a new mode of low-cost tourism, called "popular camping," was introduced. Currently, there are more than 100 such camping sites all over the country. Additionally, there are plans to make huge investments to promote international tourism, now regarded as a major economic goal. Evidently, new potentials for ecological degradation are being created.

The Political Context

Awareness of lessening environmental quality in Cuba during recent years, along with the worldwide recognition of the global environmental crisis, urged the Cuban government to include environmental issues in the country's socioeconomic planning, together with political, educational, and cultural considerations.

Thus, the Constitution of the Republic, approved in 1975, sets forth in one of its articles:

> In order to ensure the well-being of the citizens, the State and society protect nature. It is the duty of the relevant institutions, as well as of every citizen, to see to it that the waters and the atmosphere be kept clean, and that the soil, flora and fauna be protected.

After a period of review of environmental conditions, the Law on Protection of the Environment and Rational Use of Natural Resources was enacted in 1981. The law sets out the national environmental policy, identifies the specific resources covered by the law, creates a national system for organizing the pertinent work, and establishes administrative and enforcing mechanisms to guarantee its implementation.

Although this law is the most significant step in the field of environmental protection in Cuba, as it has been the main legal basis for all the relevant work done so far, in practice it was not as effective as expected, for it left the basic responsibility of protection in the hands of ministries and institutes involved in the use of the environment, in coordination with local government bodies. For instance, soil, flora, and fauna fell under the Ministry of Agriculture; fresh water, under the Institute of Hydroeconomy; marine resources, under the Ministry of Fisheries; landscapes of touristic interest, under the Institute of Tourism; and so forth, according to the natural resource involved.

Thus, an important principle of environmental management was overlooked: most often, ecological damage results from cumulative impacts of different human activities, and not from the isolated and independent operation of a single activity. Therefore, the lack of a central agency endowed with the qualification and authority necessary to project itself on the field of environment, to coordinate actions by the numerous entities involved in that activity, and to guarantee an effective control on the fulfillment of the relevant regulations was an essential drawback of the 1981 Law on the Environment.

Furthermore, the regulations that should have provided for the organization and operation of the national environment protection system set up by this law were not issued until as recently as January 1990, which considerably delayed complementary environmental legislation.

Notwithstanding these legal limitations, significant steps have been taken during the past decade to improve environmental conditions.

In this period, emphasis has been laid on water pollution control, economic utilization of sewage, improvement of eroded soils, increase in forest coverage, and restoration of areas affected by open mining or by the extraction of construction materials. Among these priorities, scientific and technical research has been directed toward seeking the adequate solutions.

Likewise, control methods and operational standards to prevent ecological damage have been worked out, and technical reports on environmental impact for all projects have been introduced as part of the decision-making process.

The setting up of environmental commissions subordinated to local governments in Cuba's 14 provinces and in most of its 169 municipalities has provided an effective device for the surveillance of environmental problems, and it is to be expected that the envisaged strengthening of such commissions will contribute to further increase their protective role.

Almost all state agencies involved with the natural environment have declared protected areas of various kinds (among them, several fauna refuges). Regrettably, since these were set up independently, such areas do not have the necessary systemic coherence. Moreover, many of them are protected only on paper or still lack technical or financial support to ensure their effective operation.

Environmental education programs attached to the various primary education subjects are being conducted, and efforts are being made to include them in the middle and higher levels of the national system of education.

Internationally, Cuba is signatory to the Convention on the Protection of the World Cultural and Natural Heritage. As to biotic resources, Cuba has recently subscribed to the Convention on the Protection of the Marine Environment of the Great Caribbean Basin and is in the process of acceding to the Convention on the International Trade of Endangered Species (CITES) of Flora and Fauna. At the same time, Cuban institutions have entered into cooperation agreements with several programs set up by international organizations, such as Man and the Biosphere (UNESCO) and the Latin American Network of Protected Areas, Flora and Fauna (FAC).

Within the overall effort made so far in the field of environmental protection, it is necessary to point out, however, that the native fauna of Cuba has been the least favored resource, despite the fact that its situation is the most critical of all. Aside from some isolated significant achievements—such as recent government measures to save the last couples of the ivory-billed woodpecker in eastern Cuba—animal diversity is under extraordinary pressure due to harmful practices still prevailing in both the private and the state sectors.

Despite numerous hunting and capturing bans issued by the agency in charge of caring for the fauna, violations are frequently committed. Among the factors favoring this situation is the fact that the corps of forest guards is obviously insufficient in both numbers and qualifications.

Illegal trade in Cuban wildlife is now increasing. Not idly we read in documents from the United Nations Environmental Program that the illegal trade in flora and fauna is perhaps the second most valuable illicit business in the world, surpassed only by narcotics.

Partly conditioned by the lack of coordination among institutional protection functions, the indigenous fauna is presently subject to various negative impacts such as the implementation of wrong silvicultural practices, or the poorly controlled release of exotic animals in natural areas (with the aim of "improving" the fauna), or the massive promotion of popular camping without the indispensable prevention of its environmental impact, among other actions contrary to the maintenance of animal diversity.

But the disadvantageous status of the fauna as compared with other natural assets in Cuba has deep historical and cultural roots that are worth examining in greater detail. In my opinion, this situation is determined by three kinds of related problems.

First, when a nation has taken shape, generation after generation, amid the plundering of its natural resources, it is understandable that people worry more about the most immediate threats, such as pollution, than they do about allegedly less significant dangers, such as loss of animal diversity. Indeed, decline in diversity is accepted as an unfortunate but unavoidable consequence of industrialization and development.

However, the most deplorable result of that historical background is the rooting out of respect and love for the fauna as an essential component of the cultural identity of any nation, and that is the case in Cuba. Besides the impoverishment of its fauna, decades of cultural colonialism have caused a great part of the Cuban people to lose any comprehension of the difference between exotic and native animals. This becomes apparent in literature, painting, and even in postage stamps currently produced in Cuba. Two of the most renowned Cuban painters have each glorified a species of introduced bird. Likewise, the white-tailed deer, responsible for significant damage to native animals after its introduction from North America in the last century, has been praised as a symbol of "Cubanness" by one of Cuba's most outstanding contemporary poets.

Another of the problems conditioning the unfavorable status of the fauna is the scarcity of scientific popularization of natural assets, which is noteworthy in view of the educational revolution and the remarkable development of the mass media that have taken place in Cuba for the past 20 years. Only quite recently has there been a rather modest response to this serious deficiency, through the publication of some books on Cuban animals, and the appearance, as yet too sporadic, of newspaper articles and radio and television programs on this subject.

The case of museums also belongs in the framework of the popular dissemination problem. For reasons not clear enough as yet, Cuba did not have a natural history museum, as an institution responsible for public instruction about nature, until the Felipe Poey Museum was opened in Havana in 1964. It was designed then merely as a set of exhibit halls with neither research collections nor staff qualified to conduct research. Later on, four other natural history museums were set up in other provinces, following the same pattern as the one in Havana. The inadequacy of these so-called natural history museums seriously impairs what should have been one of our most potent means of spreading the conservation message to our citizenry.

However, in recent years there have been some positive changes. The museum in Havana became the National Museum of Natural History, and it was entrusted with creating exhibition and education standards for the other museums of natural history. A team of 15 research workers and properly trained technicians was assigned to work as curators in the National Museum, and a similar buildup is envisaged for the other museums. Some curatorial research projects have been undertaken, and work in coordination with the National Institute of Physical Planning is under way on a program to salvage endangered species and other significant biotic elements in selected areas expected to undergo severe modification because of socioeconomic development. Educational programs are being implemented for these museums according to current needs, and one of the first methodological guidelines given to exhibit planners was to substitute the old elephants and lions for items more representative of the indigenous fauna in each locality.

Another positive result is the participation of museum staff members in natural history tourism. Every year ornithologists from the National Museum guide groups of bird-watchers visiting Cuba in winter, and experts from another museum explain to tourists the natural assets of an important national park in the eastern region, along interpretative paths surrounding the forest.

While we are very aware that much remains to be done before natural history museums realize their full potential as educational institutions, there is reason for hope.

The third kind of problem concerning the current and potential status of animal diversity in Cuba that I would like to address is the professional training of the staff involved with this natural resource.

At present, about 50 biologists work as systematic zoologists at various institutions in Cuba, most of them devoted to research applied to human, animal, or plant health. However, biology faculties do not include in their curricula the training of systematic zoologists, zoogeographers, or vertebrate paleontologists, disciplinary profiles essential to carry out faunal inventories, as well as to manage animal communities and forecast long-range effects. In the extraordinary effort the country is making to attain levels comparable to those of developed countries in the field of molecular biology, an important fact has been overlooked: most of these countries made their inventories of flora and fauna and gathered the basic data on the biotic resources necessary for development many years ago.

Owing to this deficiency, the technical work in the area of management and protection of faunal resources is usually done by professionals from other fields such as physicians, geographers, economists, and agricultural or forestry engineers, a situation that sometimes has led to unsound decisions.

This situation has much to do with the limited effectiveness attained so far by the existing fauna refuges, most of which have been based more on the draw of the so-called "charismatic megafauna" than on the recognition that the fate of animal diversity is decided by the less spectacular communities. But there is not enough basic information on the distribution and biology of most taxonomic groups. Monitoring in protected areas has barely begun, and where it is done, it has not been sufficiently sustained to measure the direction and rate of the changes. It is significant that the publication of a red-data book of endangered species, which would do so much to boost monitoring development, has not yet become a reality.

These are some of the problems to be urgently overcome. As pointed out above, legislation regulating the structure, organization, and operation of the National System for the Protection of the Environment and Rational Use of Natural Resources was issued just two months ago. This regulation sets up 11 environment protection subsystems according to the kinds of resources available. Furthermore, it creates a National State Commission as the governing body for environmental affairs, attached to the Executive Committee of the

Council of Ministers, all of which redress the operational failure of the 1981 Law on the Environment.

Likewise, a new National System of Protected Areas and the budget appropriations required for its effective operation were also approved recently. This new system of protected areas, designed with a more realistic perception of its possibilities, is made up of a national network of 55 areas and of numerous provincial networks amounting to 287 areas. The national network includes 12 natural tourist areas, 11 integral management areas, 9 fauna refuges, 8 managed flora reserves, 5 national parks, 5 natural reserves, 4 biosphere reserves, and a national marine park. All in all, the national network of protected areas covers 12.2% of the Cuban territory.

As a result of these latest actions, a promising integral environmental program is now emerging, based on ecological principles and on the fundamental concept of sustained development.

Currently, Cuba is setting itself new and pressing socioeconomic development goals, amid a devastating world economic crisis and the increasing pressure from the rich countries to pay ever less for the commodities they get from the Third World and to charge ever more for the products they sell to it.

Among other deplorable consequences, in less developed nations this state of affairs often brings about the tendency to overestimate expected economic benefits and to underestimate potential damage to nature, claimed to be inevitable for the sake of human survival.

However, in Cuba this flawed perception is being actively fought against. One of the benefits derived from the battle against poverty and ignorance in Cuba is that the country and its citizens are now well placed to engage in the worldwide fight to save and improve the environment. Under such circumstances, it is feasible to place current immediate needs and the well-being of future generations on the same footing in socioeconomic planning, the only way to successfully face the environmental challenge.

12 : Living Collections and Biodiversity

Nathan R. Flesness

There is not *a* solution to the catastrophe we are witnessing. The scale and speed of the disaster are far too great for any one solution, and there is unfortunately no precedent for 5 billion human beings suddenly sharing an enlightened vision of the future.

The best we can hope for is what may come from *many* solutions. A problem this large must become part of the agenda of as many human institutions as possible—from national governments and the World Bank, to zoological and botanic gardens, natural history museums, the Bureau of Land Management, and the family farm. If we are lucky, their combined activity will allow continuation of at least souvenirs, bits and pieces, from the existing kinds of life.

Cultural institutions such as natural history museums and zoological and botanic gardens have a special responsibility. Such facilities typically occur in the centers of the most human-dominated landscapes and are usually supported by the local community as a result of the urban public's fundamental interest and fascination with other life forms. In effect, these institutions serve as embassies, representing the interests of the diversity of life worldwide.

This paper focuses on reviewing the contributions being made by zoological gardens, mentions briefly the expanding role of botanical gardens, and in closing makes some heretical suggestions for increasing the contributions museums could make in ameliorating the biodiversity crisis.

Zoological Institutions

Zoological institutions are making contributions of several different kinds. First, a modest but growing number are acting directly to protect critical habitats internationally. The New York Zoological Society (NYZS), for example, has set a precedent that is hard to match. NYZS spends more than $3 million a year on international conservation-oriented activity. Such activity has included buying some of the unprotected land critical for the annual wildebeest migration between the Serengeti and the Masai Mara parks in Tanzania, funding protection of Middle Cay in Belize and of Punta Tambo in Argentina, instituting debt-swap-for-nature negotiations with Costa Rica, and contributing directly to the establishment of more than 75 protected areas on several continents. The Frankfurt Zoological society has a similar history and similar commitments to reserves and parks around the world. The Chicago-Brookfield Zoological Society, the Jersey Wildlife Preservation Trust, and some others are providing investment and support for additional *in situ* reserves around the world.

Encouragingly, these kinds of *in situ* commitments by zoological gardens are growing; presently a number of other institutions are in the process of creating commitments to assist Ujong Kulong Reserve in Java and black rhino reserves in several parts of Africa and to protect some of the remaining fragments of forest in southern Madagascar. We can hope that adopt-a-park may become as widespread a notion in the zoological community as adopt-an-animal.

Second, zoological gardens, like natural history museums, have a chance to affect their visitors—both intellectually and emotionally. About 120 million visitors pass through North American zoological gardens each year, so the opportunity is considerable. Modest but increasing emphasis is being given to presenting information on the origins of diversity—evolution—and the present dramatic rate of loss of diversity. Some collaborative traveling exhibits and major new visitor education centers, such as the St. Louis Zoological Gardens' new "Living World," have been created recently. This is an area where a lot more can and should be done; what seems to be missing is imaginative but practical ideas on how to do it effectively.

High-quality modern zoological exhibits give their occupants a chance to serve as ambassadors on an *affective* level—this may be their largest present contribution. As the parent of a three-year-old child, I understand that seeing, hearing, and smelling an actual elephant, a gorilla, a Sarus crane, or a gavial make an enormous impression. The interest and caring about wildlife that such exhibits can encourage are not replaceable, even by the marvelous natural history photography now available (as the film credits often indicate, close-up footage used in the best programs are often from zoological gardens).

Some of these roles are open to all museums and other cultural institutions. Museums of all kinds can make contributions to ameliorating the biodiversity crisis. However, the class of museums where the specimens are not yet dead are specially placed and becoming organized to make some unique contributions.

Botanic Gardens

Botanic gardens, under the auspices of the World Conservation Union (IUCN), have formed a conservation-coordinating body, the Botanic Gardens Conservation Secretariat (BGCS), which has carried out partial surveys of their holdings and found 29 extinct species and many more extinct varieties growing under their collective cultivation (*Extinct Plant Species of the World,* IUCN Botanic Gardens Secretariat, April 1989). This bᴏ dy has decided on restoration of the flora of the island of St. Helena as a flagship project. Here in North America the Center for Plant Conservation at Arnold Arboretum is organizing *ex situ* propagation of critically endangered North American species, in parallel with efforts to protect these last remnants of wild populations *in situ.*

Zoological Gardens

Zoological gardens have an even more complex set of species preservation programs under way. Some have quite a history; long before any talk about the "biodiversity crisis," the North American Plains bison was almost exterminated and the European bison (Wisent) went extinct in the wild (earlier in this century); both were restored to what is left of their "wilds" from zoo (and ranch) herds.

Other taxa are so far only half so lucky—the Mongolian wild horse is now in its fourteenth generation of zoo-based captive propagation and is thought to have become extinct in the wild sometime between 1947—when a single last wild horse was caught—and the 1970s. Most of what remains of the species descends from founding stock caught between 1898 and 1900. Reintroduction plans exist within the USSR, Mongolia, and the People's Republic of China, but none has yet taken place.

During the last 10 years, the new and developing science of conservation biology has emerged from a mix of evolutionary and population biology, with a critical admixture of stochastic risk assessment techniques. The results have changed perspectives on the survivability of small populations. Many of the world's protected areas are now seen to be inadequate in the short or long term to preserve in a "natural" state the flagship species they were typically established to preserve. Many remnant wild populations are so small that equilibrium genetic diversity and, presumably, future adaptive evolution are greatly restricted. The Florida panther numbers 30 or so, and the Red wolf has been successfully reintroduced into initial habitat that may support 10–20 individuals.

The Nature Conservancy, for example, is guided by studies showing that prairie remnants do not retain their invertebrate fauna when they are less than about 10 hectare in size (R. Jenkins, personal communication). Changing focus from small invertebrates to what remains of terrestrial megafauna, it is clear from both the evolving conservation biology theory and the considerable data (e.g., black rhino have been declining at 97% per generation during the twentieth Century) that we have *already* left too little protected "wild" land to conserve much of the megafauna without intensive management. The Protected Areas Data Unit of the World Conservation Monitoring Centre in Cambridge, U.K., indicates that 3.7% of the earth's land area is legally protected (J. Harrison, personal communication); the other 96.3% is either developed or available for development. Most of the protected areas that hold megafauna (i.e., the Serengeti) are becoming islands in a sea of human-dominated landscapes and are, or will be, fenced. Increasingly intensive management is required of these decreasing fragmented and isolated habitats.

The megafauna requires the most space and typically has the most economic value when harvested (or overharvested)—so it is disproportionately at risk. It is not rare for megafaunal species to disappear well before their habitats do. For example, the land surface of Europe has much charm and wildlife, but species weighing more than a few kilograms or that represent real competition or threat to man or livestock have been hard to find for hundreds of years.

Zoological gardens, somewhat by coincidence, hold a sample of life forms biased heavily toward the megafauna. In the last 20 years, a series of practical, legal, ethical, and economic factors have encouraged zoological institutions to master the breeding of most taxa. Today, 92% of new zoological mammal acquisitions are captive bred (*ISIS Species Distribution Report*, 1990).

We have, therefore, a set of institutions distributed around the world, supported by local recreational and cultural funding, with extensive holdings of increasingly endangered taxa and extensive experience in its multigenerational husbandry. At present 400 of them (about half of the planetary total) pool the vital pedigree and demographic data on their dynamic collections through one centralized computerized database—the International Species Information System (ISIS).

We thus have coherent data on the inventory, population trends, and dynamics of 150,000 live specimens, and more of their ancestors, of 4,200 taxa *ex situ*. The next step is turning this information into appropriate action. This is being done at regional and national levels by the various professional associations of zoological gardens. At the beginning of the 1980s the American Association of Zoological Parks and Aquaria (AAZPA) formally began its species survival plans (SSPs). These comprise long-term management plans linking the live collections of 150 institutions into explicit programs to preserve genetic diversity within the captive population while ensuring demographic stability.

Similar programs began in the United Kingdom at about the same time, have since arisen in Continental Europe and Australasia, are beginning in Japan, and are in early stages elsewhere. Presently 153 taxa are covered by the combination of these programs, and the presumption is that up to about 1,000 taxa can and will be so treated as we learn how to do this more efficiently. These regional programs are being coordinated globally through the Captive Breeding Specialist Group (CBSG) of the Species Survival Commission of IUCN, the World Conservation Union.

Two recent North America SSP programs of note are the California condor and black-footed ferret. In both cases, inadequate remaining habitat and unsolved wild management problems led to a remnant wild population of a dozen or so, which was captured (after great controversy). Both taxa are presently doing quite well in captivity; the high-fecundity ferrets have grown from 17 to 153 in two years and plans are in place to begin reintroductions in about 2 more years—with a minimum captive stock of 500. The condors will require more patience. They are breeding well, but they have much slower population growth rates. Meanwhile their closest relative, the Andean condor, has been reintroduced into the Andes to develop techniques, and a population of Andeans has been introduced temporarily into California, for further practice.

In the case of the ferret, its apparent extinction in the wild has led, not just to an emergency *ex situ* program by the Wyoming State Department of Game and Fish, with advice from the IUCN CBSG and now the involvement of a growing number of zoos, but also to the protection of more and more habitat as the realization grows that considerable areas will have to be made safe for ferrets and for prairie dogs, their obligate prey.

Mike Soulé (personal communication) has suggested that 2,000 vertebrate species will be lost in the next century without such emergency approaches. Zoological gardens may rescue only half of them. The loss of the other half will be tragic; many habitats *are* restorable. Restoration ecology is going to be a growth industry, and a wide diversity of techniques will surely develop. Many megavertebrates can be restored by economic rather than ecological changes. The Arabian oryx has now been restored to the Omani desert from zoo stock where it spent 20 years in escape from a temporarily lethal habitat. The habitat has now been made tenable by hiring the Bedouin, who used to poach the oryx, to protect it.

In short, captive propagation is not the ideal approach to conservation, but it is already late in the day and there are a great many taxa that will be lost without it and similar intensive management techniques—whether applied in increasingly natural and scientific zoo settings or in increasingly managed and confining parks and reserves. The 500, 1,000, or perhaps more species that may depend on the role assumed by zoological gardens are a pitiful and biased fraction of what is being lost, but they are far better than nothing. Zoological gardens are primarily locally funded, municipally based organizations. There is no precedent for such institutions' assuming responsibility for a portion of planetary life, but along with botanical gardens, they are doing so. We hope other institutions can find ways to link the interests of their supporters with what needs to be done and to contribute some different small pieces of the solution.

Expanding the Contributions of Museum-Based Science

I have a few specific suggestions on ways natural history museums could make a larger contribution to the solutions that are evolving.

1. Some conservation researchers have decided to accept the common museum claim that museum collections are valuable documentation of fast-vanishing diversity. For example, the International Council for Bird Preservation (ICBP) is building a database on the 3,000 or so restricted-range birds of the world, assembling information on dates and sites of occurrence primarily by sending their staff around to the world's museum collections to read the tags on the birds (T. Johnson, personal communication). I cannot think of a way that this possibly valuable information could be made *less* accessible to those who may be able to use it for conservation action purposes. A project to make this information available to conservation-oriented researchers and planners would be of value.

2. Museums support considerable systematics activity. Fortunately, many of the last-moment species rescue activities based in zoos do not need much systematics information—there is often only one small local population left. However, in the increasing number of cases in which a decision is taken to intervene earlier, the programs are desperate for guidance on the boundaries of the evolutionarily significant units that may be involved. In general, of course, for taxa like gorillas, chimpanzees, rhinos, elephants, no contemporary molecular or other discrete-bit information-rich systematics work is available at the time intervention must be accomplished. My point here is that there are an infinite number of potential systematics stud-

ies that researchers *could* choose to perform next—what about focusing more of the limited systematics resources on taxa for which we must make irreversible management decisions, both in the field and in captivity, in short order?

3. Lastly (and most heretically), conservation programs of all types are consumers of the taxonomic products produced by the systematics community. The Linnaean scheme has one critical flaw—the generic assignment of a taxon is stored in the *name* of the taxon. This fundamental confusion of label with relationship offers great opportunity for development of specialist knowledge *about names.* However, time and resources are short. We need to be investing instead in knowledge about organisms and their ecosystems. At the World Conservation Monitoring Centre (WCMC), more than half of the staff resources available for assembling global overviews of the plant kingdom were invested in resolving name confusions (WCMC, personal communication, 1990). In this age of computer tools, I should like to challenge the museum portion of the systematics community to reexamine the products you offer us consumers and provide services that would more effectively *support* conservation of the diversity the systematists are trying to describe. The requirements are stability of the *labels,* delivery to the public domain of both the names and synonymy information in *machine-readable form* for wide groups of organisms, and flexible systems for categorizing the taxon's *rank* separately from its label. I fully well recognize the obstacles to such a suggestion but ask that you consider as well the magnitude of the benefits.

REFERENCE

Isis Species Distribution Report. 1990. International Species Information System. 12101 Johnny Cake Ridge Road. Apple Valley, Minnesota 55124.

13 : Third World Museums and Biodiversity

P. E. Vanzolini

My main topic is the role of zoological museums of tropical, underdeveloped countries, in the study, monitoring, and conservation of biodiversity. This is, of course, the prime rationale and excuse for the existence of such museums and has been my personal and institutional concern for the last 44 years as a practicing and undeviating museum curator and systematist. What I have to present is a distillation of this professional experience and concern, both from the viewpoint of museums in general and from that of my country, with its peculiarities and limitations, all tempered with the search for a theoretical, evolutionary, and ecological goal that is the common denominator of research in natural history.

I start with a brief historical outline leading to the operational concept of what we in the profession understand by "biodiversity." The scope of the subject has made it impossible to avoid a measure of digression; practitioners will understand. I have been also unable to avoid (my reviewers say) some sweeping statements. Let it be fully understood that they sweep no wider than the field of vertebrate zoology in tropical lowland South America.

After presenting my view of what should be studied and preserved in the context of diversity, I look for ecological and evolutionary parameters suited to the framing of general museum policies.

For our purposes, a consideration of the development of the concept of biotic diversity must start, as a matter of course, with the second half of the eighteenth century, i.e., with the mature works of Linnaeus.

In these present times of estrangement between the natural sciences and the humanities, the prestige of Linnaeus has been at a quite undeserved low ebb. People are wont to see (not quite to read) the *Systema Naturae*, (Linnaeus, 1758) with its synthetic, integrated diagnoses, and judge it not as the classic it is—the close of a cycle of much acquisition and systematization of knowledge—but by the standards of modern monographs. The *Systema* cannot be properly understood without reference to the numerous works cited in it and, especially, to Linnaeus' "pre-Linnaean" (pre-1758) works, such as *Museum Regis* (1754) and parts of *Amoenitates Academicae* (1749–1769), which contain much detailed description and comment and do not fit the format of a system of the three kingdoms of Nature.

In any case, regardless of current ideas about the *Systema,* it was consistently acclaimed by its public as providing the first framework for the development of new scientific concepts, in the fields, not only of systematics, but also of biogeography, as a consequence of its concern with the "origin of species," or evolution. This concern with multiplication of species is, and has been, at the core of systematic thought and thus of museum research.

The raw materials of the *Systema* and its supporting works were mostly acquired in the preceding century, i.e., when there was already marked interest in, if not yet much documentation of, the "natural products" of parts of the world other than Europe. Witness the broad use by Linnaeus of works mostly, if not entirely, concerned with exotic faunas, such as Seba's (1734–1766) or Marcgraf's (1648) *Historia Naturalis Brasiliae.*

The *Systema* in fact permitted its own continuation and updating and allowed a first view of the world fauna as a whole. This awareness of geographic diversity could not help but be incipient and somewhat naive. For example, the distribution of a single species encompassing Asia and America was not thought to be strange—zoogeography had not yet come to the aid of systematics.

Thus, the next impulse in the evolution of the concept of diversity came with the systematized scientific exploration of parts of the world distant from Europe, initially with the cycle of long sea voyages, made for discovery and charting. This customarily entailed natural history (which included anthropology), collections and field observations being made by ship's surgeons or, and progressively more so, by professional (or intended, such as Darwin) naturalists (Vanzolini 1977–1978:51, 58, 59, 61, 73, 78, 81, 92).

The great books that resulted, usually known from the names of the exploring ships, completely changed the outlook of the natural sciences in the nineteenth century, adding decided geographic dimensions to the purely taxonomic concept of a "systema." The next (actually overlapping) step came naturally, as the purposeful, planned exploration by individual expeditions of the natural history of the less known parts of the world, especially the "tropics," whose major attractions were, from the beginning, strangeness and diversity, real or imaginary. It is the time, for us South Americans, of Humboldt (see Feisst 1978), of Spix and Martius (1823), of Prince Max (Wied-Neuwied 1820–1821, 1825). From this period and this type of expedition also resulted great books, which not only strongly contributed to the development of the systematics of individual groups but also were decisive in framing the concept of regional tropical faunas—or as one might say, of diversity.

This is also the time when emphasis passed from the individually owned cabinets of curiosities to scientifically organized national museums. This change of stress was accompanied by important conceptual ones. There was a growing, and finally an overwhelming, preoccupation with the completeness of collections, in part, it is true, a hangover from the vanities of the aficionado days, but in part an authentic awareness of the diversity of faunas. Particularly illustrative are the successive "editions" (in fact, new works) of the British Museum (Natural History) catalogues. These are treatises on the respective animal groups, always the best to be had, at their time and long after. They all characteristically start with a careful accounting of the number of species then known and of those owned by the Museum.

There are several interesting early by-products of this philosophy of collections. One is the recruiting of nationals living abroad (not only in the colonies) to collect for the mother country's museums. All European countries engaged enthusiastically in this pursuit; at that time of active emigration, many of them were signally successful: England, France, Italy, Germany, Holland, and even the Scandinavian countries and Switzerland. These citizen-contributed collections in many cases turned out to be preeminent, in regard both to the importance of the holdings and to the published results of the investigations. An impressive case is that of Peter Wilhelm Lund (Stangerup 1984), who, having moved to Brasil for his health, ended by revolutionizing Pleistocene paleontology and zoogeography and by causing invaluable research to be published on all aspects of vertebrate zoology. (I stress the "published" as the only criterion of appraisal from the viewpoint of the host country.)

Conversely, an undesirable by-product of collecting by the diaspora was, from the beginning, and still is, the production of much incompetent work in institutions enriched by gifts of occasional specimens from odd corners of the world (such as Brasil) but unprepared for the task. This type of research, especially the inadequate descriptions of new species, can originate, and often did, unwelcome impediments to research in the underdeveloped countries, especially when access to the actual specimens is not easy. This frequent issue legitimately lies at the root of many "nationalistic" restrictions to collecting by foreigners.

Finally, the need to encompass diversity in the collections led to the policy of exchanges and loans that is one of the most characteristic and attractive of museum policies, at present more important than ever.

Returning to the main argument, the accumulation in museums of collections made and studied with a geographical bias resulted, at the turn of the 20th Century, in the first of the great subspecies movements, a revolution that ultimately became frustrated in itself but that eventually developed into modern systematics.

It all started with museum lepidopterists and ornithologists (Vanzolini 1986b). In the late years of the nineteenth and in the first quarter of the twentieth centuries, they found themselves in a very favorable position. Their materials were dry specimens, easy to prepare and convenient to store. These subjects are highly visual animals, so that related forms are ordinarily distinguished by details of color pattern, whose intermediate states are quite obvious to the naked eye and require no computations. Having good coverage of several large areas of the world, European (mostly German) ornithologists, for instance, did very precocious and creditable work on bird geographic speciation, some of which is really classic and quite unjustly neglected nowadays (Vanzolini 1986b).

The concept of subspecies, then originated, is explicitly or implicitly interwoven with the development or succeeding concepts of diversity and has become a major field of activity in museums.

It started straightforward and clear-cut: a population occupying a definite geographic area, throughout which it is morphologically uniform and at the edges of which it intergrades over short distances with similar populations occupying adjacent areas. In the application, however, the concept underwent some vexing shifts, which I believe originated in part from the unfortunate term *subspecies,* of regrettable nomenclatorial descent. The subspecies being ranked nomenclatorially below the species, this subordination permeated the whole concept, and subspecies came to be widely, implicitly, and often openly considered as, instead of geographical races genetically compatible, merely as taxa differing among themselves by features "less important" than "specific characters." This posture is still encountered in contemporary work.

Another common deformation of the initial concept of subspecies among zoologists was to consider them as "incipient species," which is an unjustifiable infusion of evolutionary content into a purely morphological concept. Contrariwise, it has become clear that many, perhaps most, subspecies are in fact aborted species that failed to reach genetic identity during a period of isolation.

Parenthetically, there has been another, but short-lived, deliberate transmogrification of the subspecies concept. In some fields of zoology, especially entomology, riddled with amateurs proud of their collections, there existed the practice of naming, in Latin and with intended nomenclatorial validity, "varieties," whose types, with their red labels, graced the collections of the describers. When varieties as a nomenclatorial level were rejected by the International Commission on Zoological Nomenclature, some authors continued to describe them, now styled as "subspecies," even to the extreme of contradictorily describing more than one from a same locality. As said, this trend was fortunately short-lived, pruned out by journal editors, and need not concern us here.

The next phase in the evolution of the argument was the movement formally initiated by *The New Systematics,* edited by Sir Julian Huxley in 1940 and propelled into full strength by Dobzhansky's (1937) *Genetics and the Origin of Species* and Mayr's (1942) sister volume *Systematics and the Origin of Species.* These, and especially the latter, systematized and summarized the biological basis of the subspecies concept, whose canon they established, and incidentally initiated a new cycle of deformations. Of these the main one was actually not of concept but of application: systematists tended to consider as subspecies any morphologically similar allopatric populations, skipping the verification of a "hybrid zone" (genetic jargon for an intergradation belt). This trend is particularly evident in the practically automatic, unsubstantiated ascription of subspecific rank to insular populations.

This cavalier attitude attracted much criticism; a very useful consequence was the realization that many allopatric, morphologically similar forms were actually good species, not genetically connected geographical races.

In spite of much work done by fine geneticists and even systematists, present applications of the subspecies concept are uneven, frequently undocumented, and lead to no improvement of either evolutionary theory or practical taxonomy. The upgrading of this field of research is an obvious museum task.

The ongoing phase of the concept of biodiversity is characterized by the application of genetic (karyotypic and molecular) concepts and techniques to natural populations. The relationships between genetics and morphology, previously taken, in the Dobzhansky and Mayr tradition, to be straightforward, even if not always immediately obvious, are at present going through a period of reappraisal. Technical improvement and simplifications are progressively making possible adequate geographic mapping of genetic features. Great strides are being made every day and there is every reason to expect much more in the near future. An uneasiness remains, though: the frequent divorce of modern techniques from basic morphological work. To this point I shall return.

In parallel, but not always in synchronism, with the attempts of zoologists and geneticists to understand speciation mechanisms, there has been a long-standing ecological preoccupation with the problem. This has developed along two main lines. A first one has dealt with the origin, maintenance, and measurement of diversity at the ecosystem and community levels (Magurran 1988). It has remained a firm province of ecology. On the other hand, the branch of ecological zoogeography attempted to analyze distributions and differentiation on a continent-wide scale. This was for a long time a dull, descriptive field, until it received the input of paleoclimatic data. These added to the zoological studies the dimension of time, plus increasingly reliable paleoecological information. There is nowadays a clear trend toward the consolidation of the disciplines involved into one interdisciplinary whole: the object of research becomes no longer one form or one community but an individualized complex of biological interactions, the outcome from an active past in a particular geographic and ecological context.

I hope it has by now become clear that diversity, as faced by a museum zoologist, cannot be taken as only an array of sympatric species, detached from their ecological and evolutionary contexts: interactions and even mechanisms of a higher order are an inseparable part of the ensemble, and their study and preservation are specific problems for consideration (Vanzolini 1980). Exclusive emphasis on species (especially "threatened"), or on arrays of species, is a common failing of conservationists à outrance; to me it is equivalent to refusing to understand the nature of what one is trying to conserve, in favor of a strictly "hands off" policy, based on emotion and lack of culture.

Once we thus come to consider diversity to include not only the existing results of evolution but also the respective processes, we have to take time and space objectively into account.

It was long believed that evolutionary processes occurred in "geologic time," i.e., on a time scale of millions of years. As P. J. Darlington Jr. once remarked, dating by carbon 14 caused one of the most drastic revolutions in any field of science, a reduction of the time scale by a factor of 10, to the order of magnitude of merely thousands of years. Laboratory work has in fact shown that evolutionary processes can be considerably fast; chance observations have revealed some natural examples of very rapid change. In my own area of work there are two impressive cases.

Bothrops insularis is a pit viper occurring on one uninhabited island on the coast of the state of São Paulo. It was discovered and described in the early 1920s (Amaral 1922) as a risk to the life of lighthouse keepers. The population has since been sampled a few times, and in 1954 it was discovered that a high proportion of the individuals lately collected were intersexes. Reference to the old materials in the systematic collection permitted a study of the evolution of this condition, which had been extremely quick, and its importance is evident (Hoge, Belluomini, and Schreiber 1954; Hoge et al. 1961).

This example, besides its fascinating scientific aspects, raises the very important question of the ethics of the hands-off policy, which takes for granted—even in the usual absence of quantitative data—that any interference with natural populations is detrimental. It may be questioned whether such an ideal risk affords reason enough to block the acquisition of knowledge, including knowledge that may be essential to the conservation of the species or the community.

Another example of fast evolutive change, especially interesting on account of its serendipity, happened to me. In 1965 an airline, as they are wont to do, left me stranded for a few days without my equipment in the city of Obidos, on the northern bank of the Amazon. Following museum routine I made myself a slingshot and went collecting lizards. Of one species I collected two males and three females. Three years later I went back to Obidos, this time on purpose: the species had been discovered to have both bisexual and unisexual (parthenogenetic) populations, and I was rechecking its distribution. I collected 108 specimens, all females (Vanzolini 1970). Fast change indeed.

The interest of longitudinal collecting is obvious; how it should be carried on is not. While it is clear that sampling definite localities at given or even at irregular intervals might be useful, it is clearer still that this cannot be made a blanket recommendation—its widespread application would stretch museum resources at all levels beyond the imaginable. But the point should be kept in mind by museum zoologists, in the context of individual problems, or localities, without ever forgetting serendipity; if one locality, especially one well explored the first time, can be explored again, after a reasonable interval, and without undue strain, by all means do it again, even if there is no a priori evidence of particularly enticing events.

We get now to the point that is for me the crux of the argument: the sampling of geographical and ecological space, from the viewpoints of research to be initiated, of documentation for its own sake, and of preservation of biological evidence.

First of all a rationale is needed. Theoretically a "shotgun" approach would be conceivable: cover the landscape as closely as possible with a network of collecting stations. Not only would the area be well covered; this type of sampling would permit the drawing of valid isophenes, at one time such an attractive concept. Unfortunately, for the areas with which we deal, of continental size or one order of magnitude below, the scheme is not feasible. Thus material restraints lead us to look for a rationale to organize the space. In the case of South America, two schemes have been proposed: Ab 'Saber's (1977) morphoclimatic Domains and Holdridge's (1947) life zones.

The latter, based on the joint consideration of latitude, altitude, and rainfall, is enthusiastically adopted by ecologists working on relatively small areas of strong relief (e.g., Janzen 1983). There was even at one time some pressure from international agencies to have it uniformly adopted in tropical

South America. In this area, however, the scheme turned out to be meaningless, for altitude and latitude are bad ecological proxies there, and rainfall tolerances are sometimes strikingly broad.

Ab 'Saber's model had as a starting point the pragmatic notion that in Brasil some characteristic types of natural landscape, occupying ample areas, were consensually recognized and unambiguously named. The model was developed in two steps: first, to identify the factors that determined such types of landscape with a broad continuous distribution, and second, to use these criteria to map the morphoclimatic Domains. Five environmental features ended by being used to define the Domains: relief, climate, soil, drainage pattern, and vegetation. Of course, there is full recognition that these features are not mutually independent; on the contrary, their interactions are a fundamental if implicit part of the model.

Ab 'Saber's morphoclimatic Domains have a polygonal core area, of subcontinental dimensions (hundreds of thousands to millions of square kilometers), characterized by the superimposition of given states of the criterion features. For example, the core of the hylaea, the area of the Amazon forest, is thus individualized: (1) the *relief* is one of large alluvial plains or low plateaus; (2) the *climate* is isothermal, superhumid; (3) *soils* are diversified but in general shallow and poor in exchangeable bases; (4) *drainage* is mostly allochthonous and poorly ranked (i.e., the major rivers arise outside the Domain; there is not a definite hierarchy of tributaries, such as one finds in dendritic drainages); (5) the *vegetation* is the equatorial rain forest.

For comparison, an adjacent and contrasting Domain, that of the Central-Brasilian cerrados, is thus defined: (1) relief of rolling or mesalike low-altitude plateaus; (2) climate with a cool, dry winter (otherwise no extremes of temperature), rainfall at subhumid or even humid levels; (3) very deep soils, functioning as a capacious water reservoir, chemically poor and acid; (4) dendritic, perennial drainage, with gallery forests; (5) open vegetation of a peculiar, well-defined type ("cerrado"), evergreen, with extremely deep root systems, and with very thick bark but no adaptations, physiological or morphological, against water loss.

This scheme, as said, corresponds to and to a large extent explains landscape types traditionally recognized in Brasil. Its extension to the remainder of lowland South America is easy, logical, and rewarding. Not so in what concerns the Andes, where altitude is, of course, fundamental and precludes extensive areas of continuous ecology. I limit the application of the model to cis-Andean South America, which introduces the further consideration that the concern of present museum zoologists with diversity can hardly be world-wide or even continent-wide but has to be geographically limited in scope. The specification of each area of competence tends to be a compromise between zoogeographic criteria, the realities of nationalism, and the availability of funds.

One extremely important feature of the model of morphoclimatic Domains, and one that makes it, for our purposes, much superior to others, such as biotic provinces, is that the core areas are not pressed against one another like tiles on a floor or countries on a map but are separated by belts where one or more of the characteristic Domain features are lacking. These belts may be very broad; their ecological features usually vary much from place to place. They may have ecological aspects intermediate between the cores, or show mosaics of contrasting environments, or even reveal special, individual characteristics. No specific field work has yet been done on them, but it is to be expected that a wealth of evolutionary opportunities may be open for the faunas of contrasting adjacent Domains.

Another very important feature of the model is that paleoclimatic evidence from various sources, but mainly geomorphological, indicates that they are potentially very mobile and that they have responded to climatic alternatives by successive expansion and retraction of the core areas. This means that, at least since the establishment of the modern floras, essentially the same types of Domains have existed, but with very diverse extent and arrangement, according to the prevailing world climate. The most drastic climatic alternations, also the most visible, because so recent, are those connected with Pleistocene glaciations, and in fact the model of Domains has been employed (as I see it, successfully) to recent speciation events believably related to vegetational changes (e.g., Whitmore and Prance 1987).

One important detail of this issue of pulsations of core areas during climatic cycles is that retreat is not always total: that very frequently "islands" of one type of environment are left behind in favorable spots in the interior of the contrasting, reexpanded core area. There are thus, for example, numerous enclaves of open formations, of diverse sizes, inside Amazonia; conversely, there are spots of forest in favorable (mostly orographic) situations in the semiarid Domain of the "caatingas" of northeastern Brazil. These enclaves are natural laboratories where experiments on geographical speciation are being run—in fact several of these experiments have been identified and studied (e.g., Vanzolini 1986a).

I believe a rational strategy of sampling based on the model of morphoclimatic Domains, although extensive (but such is the continent) and expensive, is feasible, much easier to put into effect than an eventual shotgun approach. It would be, in fact, a simple extension of strategies already in use in zoological research in cis-Andean South America.

Keeping in mind that we are dealing, not with the preservation of a well-known and cherished heritage, but rather with the materials of our current research and that we discover as we go, such a scheme would follow several lines of thought.

One includes the methodical sampling of the core areas. It has been mentioned that core areas have been, during climatic cycles, alternately dissected and reassembled. The evolutionary events consequent to these shifts are important and should be analyzed as meticulously as possible, for they are likely to furnish guidelines for advancing research and for orienting conservation. In fact, a breakthrough paper on tropical speciation, Haffer's (1969) study of patterns of bird differentiation in the hylaea, started from the null hypothesis that to a uniform (at the level of the problem) ecology should correspond uniform patterns of distribution. The null hypothesis, it is well known, was rejected.

Any scheme of the preservation of diversity including the preservation of processes has to contemplate conservation in the field, not only in the museum, i.e., the constitution of reserves, not untouchable, but to be tapped, on occasion and with scruple, for the scientific information not available on the collection shelves. The model of morphoclimatic Domains in its museum version affords reasonable guidelines.

Preservation of significant fragments of core areas is theoretically a simple problem. Preservation of a terrestrial ecosystem as a whole hinges on the maintenance of viable populations of the top predators. Theoretically, again, this is not too hard a monographic task. In practice, however, no studies of top predators are available, and in spite of some talk to the contrary, the only strategy so far adopted is an opportunistic one: grab and preserve as much as you can, wherever and whenever you can. Such an attitude is frequently unavoidable, but when possible, the fragments to preserve should be selected on the basis of existing studies of geographical differentiation. In the absence of these, they might be distributed according to eventual natural subdivision of the landscape.

It was at one time proposed in Brasil that special attention be given to sections of core areas thought to have been unaffected during the unfavorable leg of climatic cycles ("refuges"), for they would encompass maximum diversity, in the sense of number of species extant. The viewpoint I defend (Vanzolini 1980) is, of course, that other things besides maximum number of species should be conserved. Such former refuges should be part of the areas preserved, but these should also include intervening belts.

The study of enclaves is both extremely urgent, as many are endangered by encroachment, and rewarding. These are also choice sites for periodical collecting, for it is foreseeable that many of the processes occurring there are rapid, possibly detectable in the time frame of a single naturalist (Vanzolini 1981).

Enclaves also raise an important problem with regard to reserves. Small and unique, they are obvious candidates for preservation. In fact, many present enclaves are the result of preservation, small areas of pristine ecology saved in the middle of anthropic devastation.

This should imply, again, that not only the species but also the scientific information the area contains should be preserved for use whenever opportune. The estimation of the eventual damage caused by collecting or experimentation on such limited populations depends on accurate ecological knowledge of the ecosystems, again not available. Acquisition of this type of information should have a high priority. I stress that rational use of enclaves is essential to evolutionary theory.

Finally, within the framework of the model of morphoclimatic Domains, one must consider the intervening strips that separate core areas. These are belts, sometimes called "of ecological tension," where different ecologies maintain contact, even during the vicissitudes of climatic cycles. If interpecific competition exists, this should be a favorable place to look for it. Additionally, they are internally much diversified. Sampling them calls for a strategy of transects, to permit the establishment of correlations between morphological or physiological features of the animals and the spatial parameters; in other words, "clines" in the original meaning of the term (Huxley 1939). The same orientation should be used in the design of reserves, which should include segments of the core areas connected by adequate corridors across the intervening belts. The conclusion is unavoidable that, while the realities of life may force opportunism on conservation programs, it is possible and, of course, desirable to introduce a measure of science.

We turn at last to cytogenetic and biochemical methods as applied in research on speciation and diversity. They are at the technological frontier, and from them are expected the next significant steps in evolutionary theory at this level. Their use being, by nature, restricted to few specimens at each turn, they should be carefully coordinated with simple morphological methods capable of handling a large body of information and of bringing into focus problems or aspects of problems that can be conveniently solved. In other words, genetic methods applied to groups that have not yet been submitted to full traditional taxonomic treatment are liable to produce most interesting karyotypes or gels but to contribute very little to evolutionary studies, possibly even adding to the confusion. There has been a delusion, for some time widespread but now gone, that animals hard to identify by traditional methods ("sibling species") would be easily determined by means of their karyotypes.

There is in research on differentiation an obvious hierarchy of methods, from those demanding only visual inspection of preserved specimens, through the application of laboratory techniques to collection materials, to methods demanding special care in the field and elaborate techniques in the laboratory. The scope, both in number of individuals and of area covered decreases with each step; it thus seems advisable to apply these methods sequentially, following the hierarchy, each step helping to define the next, to focus the questions to be asked and to provide the answers.

Techniques are beginning to become available to extract biochemical information from routinely preserved specimens (e.g., Kocher et al. 1989). The vistas are staggering. In general, however, at present, special techniques of preservation are necessary, or live animals may be indispensable. Museums should equip themselves to properly obtain and store samples, especially from species rare or hard to collect, and should help to locate, plan, establish, and maintain reservoirs of live animals. This last item implies, of course, geographical specialization and increased collaboration among sister institutions.

Museums have always been involved with the study of diversity, especially from the viewpoint of its origin and geographical differentiation; the present generalized concern with the conservation of diversity has been a museum constant. The documentation they keep and the research they do are essential to the understanding of evolution at the species level, and this understanding is, in its turn, essential to the preservation of diversity.

The development of ecological concepts and the application of genetic techniques are the most promising aspects of modern research on speciation; they need to be tightly coupled with basic morphological investigation. On the other hand, these new concepts and techniques demand that much evidence be kept alive, for use when possible or necessary, which requires a closer participation of museums in the establishment and use of wildlife reserves.

Geographical specialization places the museums of tropical countries in the forefront; they need to perfect themselves in the traditional tasks and to get ready for the new strategies in development.

REFERENCES

Ab 'Saber, A. N. 1977. Os domínios morfoclimáticos na América do Sul. Primeira aproximacão. *Geomorfologia* (Instituto de Geografia, Universidade de de São Paulo) 52:1–21.

Amaral, A. 1922. Contribuicão para o conhecimento dos ophidios do Brasil. A. Parte 1. Quatro novas especies de serpentes brasileiras. *Annex Memorias Instituto. Butantan 1* (1) (1921):1–38.

Dobzhansky, T. 1937. *Genetics and the Origin of Species.* New York: Columbia University Press.

Feisst, W. 1978. *Alexander von Humboldt 1769–1859.* Wuppertal: Dr. Wolfgang Schwarze Verlag.

Haffer, J. 1969. Speciation in Amazonian forest birds. *Science* 165:131–137.

Hoge, A. R., R. Belluomini, and G. Schreiber. 1954. Intersexuality in a highly isolated population of snakes. *Caryologia* 6 (Suppl.):964–965.

Hoge, A. R., R. Belluomini, G. Schreiber, and A. M. Penha. 1961. Anomalias sexuais em *Bothrops insularis* (Amaral) 1921. Análise estatística da terceira amostra, desdobramento e comparacões com as duas amostras anteriores. *Annais Academia. Brasileira Cienças.* 33:259–265.

Holdridge, L. R. 1947. Determination of world plant formations from simple climatic data. *Science* 105:367–368.

Huxley, J. S. 1939. Clines, an auxiliary method in taxonomy. *Bijdragen tot de Dierkunde* 27:491–520.

Huxley, J. S., ed. 1940. *The New Systematics.* London: Oxford University Press.

Kocher, T. D., W. K. Thomas, A. Meyer, S. V. Edwards, S. Paabo, F. X. Villablanca, and A. C. Wilson. 1989. Dynamics of mitochondrial DNA evolution in animals: amplification and sequencing with conserved primers. *Proceedings of the National Academy of Sciences U.S.A.* 86:6196–6200.

Janzen, D. H. ed. 1983. *Costa Rican Natural History.* Chicago and London: University of Chicago Press.

Linnaeus, C. 1749–1769. *Amoenitates Academicae,* 7 vols. Holmiae et Lipsiae: Godofredus Kiesewetter.

Linnaeus, C. 1754. *Museum Regis Adolphi Friderici.* Holmiae

Linnaeus, C. 1758. *Systema Naturae per regna tria.* Editio decima, reformata, 2 vols. Holmiae: Laurentius Salvius.

Magurran, A. E. 1988. *Ecological Diversity and Its Measurement.* London and Sydney: Croom Helm.

Marcgraf, J. 1648. *G. Marcgravii . . . Historia rerum naturalium Brasiliae,* libri octo. Lugdunium Batavorum: F. Hackius; Amstelodam: L. Elzevirius.

Mayr, E. 1942. *Systematics and the Origin of Species.* New York: Columbia University Press.

Seba, A. 1734–1766. *Locupletissimi rerum naturalium thesauri,* . . . 4 vols. Anstelodami: Janssonio Waesbergios et J. Wetsten.

Spix, J. B. von and C. F. P. von Martius. 1823. *Reise in Brasilien . . . in den Jahren 1817 bis 1820,* 3 vols., 1 atlas. München: M. Lindauer, I. J. Lentner, C. F. P. Martius; Leipzig: F. Flaischer (facsimile reprint, 1966, Stuttgart: Brockhaus).

Stangerup, H. 1984. *The Road to Lagoa Santa.* London and New York: Marion Boyars.

Vanzolini, P. E. 1970. Unisexual *Cnemidophorus lemniscatus* in the Amazonas valley: a preliminary note. *Papéis Avulsos Zoologica,* São Paulo 23(7):63–68.

Vanzolini, P. E. 1977–1978. *An Annotated Bibliography of the Land and Freshwater Reptiles of South America (1758–1975),* 2 vols. São Paulo: Museu de Zoologia da Universidade de São Paulo.

Vanzolini, P. E. 1980. Algumas questões ecológicas ligadas à conservacão da natureza no Brasil. *Inter-facies* (Escritos e Documentos) (Universidade Estado Paulista, S. José do Rio Preto) 21:1–23.

Vanzolini, P. E., 1981. A quasi-historical approach to the natural history of the differentiation of reptiles in tropical geographical isolates. *Papéis Avulsos* Zoologia, São Paulo 34(19):189–204.

Vanzolini, P. E., 1986a. Levantamento herpetológico da área do estado de Rondônia sob a influência da rodovia BR-364. Brasilia: CNPq (Conselho Nacional de Desenvolvimento Científico e Tecnológico), Programa Polonoroeste, Subprograma Ecologia Animal, Relatório de Pesquisa 1:1–50.

Vanzolini, P. E. 1986b. Paleoclimas e especiacão em animais da América do Sul tropical. *ABEQUA* (Associação Brasileira Estudos Quaternicos.) *Publ. Avulsa* 1:1–35.

Whitmore, T. C. and G. T. Prance, eds. 1987. *Biogeography and Quaternary History in Tropical America*. Oxford: Clarendon Press.

Wied-Neuwied, M., Prinz zu. 1820–1821. *Reise nach Brasilien in den Jahren 1815 bis 1817*, 2 vols. Frankfurt: H. L. Bronner.

Wied-Neuwied, M., Prinz zu. 1825. *Beitraege zur Naturgeschichte von Brasilien*, 4 vols. Weimar: Gr. H. S. priv. Landes-Industrie-Comptoirs.

Contributors

George F. Barrowclough
Department of Ornithology
American Museum of Natural History
Central Park West at 79th Street
New York, New York 10024

Joel Cracraft
Department of Anatomy and Cell
  Biology, and Department of
  Biological Sciences
University of Illinois
P.O. Box 6998
Chicago, Illinois 60680

Niles Eldredge
Department of Invertebrates
American Museum of Natural History
Central Park West at 79th Street
New York, New York 10024

Nathan R. Flesness
International Species Information
  System
Minnesota Zoological Garden
Apple Valley, Minnesota 55124

Michael J. Novacek
Vice President and Dean of Science
  and Department of Vertebrate
  Paleontology
American Museum of Natural History
Central Park West at 79th Street
New York, New York 10024

Norman I. Platnick
Department of Entomology
American Museum of Natural History
Central Park West at 79th Street
New York, New York 10024

J. John Sepkoski, Jr.
Department of the Geophysical Sciences
University of Chicago
5734 South Ellis Avenue
Chicago, Illinois 60637

Gilberto Silva Taboada
Museo Nacional de Historia
Natural "Felipe Poey"
La Habana, Cuba

George Stevens
Department of Biology
University of New Mexico
Albuquerque, New Mexico 87131

Melanie L. J. Stiassny
Department of Herpetology and
   Ichthyology
American Museum of Natural History
Central Park West at 79th Street
New York, New York 10024

Ian Tattersall
Department of Anthropology
American Museum of Natural History
Central Park West at 79th Street
New York, New York 10024

P. E. Vanzolini
Director
Museu de Zoologia da Universidade
de São Paulo
Caixa Postal 7172
01051 São Paulo, Brasil

Judith E. Winston
Department of Invertebrates
American Museum of Natural History
Central Park West at 79th Street
New York, New York 10024

Index

Words in *italics* indicate illustrations, charts, or tables